Contributions from Science Education Research

Volume 17

Contributions from Science Education Research is the international, multidisciplinary book series of the European Science Education Research Association (ESERA). The aim of the series is to synthesize, for the benefit of the scholarly community, the findings of high quality, theoretically-framed research in the domain of science education as well as comprehensive explorations of specific methodological strands in science education research. The series aims to publish books that are innovative in attempting to forge new ways of representing emergent knowledge in the field. The series includes edited collections of chapters, monographs and handbooks that are evaluated on the basis of originality, scientific rigor and significance for science education research. The book series is intended to focus mainly on work carried out in Europe. However, contributions from researchers affiliated with non-European institutions and non-members of the European Science Education Research Association are welcomed. The series is designed to appeal to a wide audience of researchers and post-graduate students in science education. Book proposals for this series may be submitted to the Publishing Editor: Claudia Acuna E-mail: Claudia.Acuna@springer.com

Nicole Graulich • Michal Haskel-Ittah
Rayendra Wahyu Bachtiar
Editors

Exploring Mechanistic Reasoning in Science Education

Understanding Phenomena

Springer

Editors
Nicole Graulich
Justus-Liebig University Giessen
Giessen, Germany

Michal Haskel-Ittah
Weizmann Institute of Science
Rehovot, Israel

Rayendra Wahyu Bachtiar
University of Jember
Jember, Indonesia

ISSN 2213-3623 ISSN 2213-3631 (electronic)
Contributions from Science Education Research
ISBN 978-3-032-10245-4 ISBN 978-3-032-10246-1 (eBook)
https://doi.org/10.1007/978-3-032-10246-1

This Springer imprint is published by the registered company Springer Nature Switzerland AG
The registered company address is: Gewerbestrasse 11, 6330 Cham, Switzerland

Foreword: Why Mechanistic Reasoning Matters: A Perspective from the Philosophy of Science

Science educators have often argued that the most important outcome of science education is that students should have an understanding of the Nature of Science (NOS) (Holbrook, 2010; NSTA Board of Directors, 2020). While no one doubts that courses in biology, chemistry, and physics should teach students something about the major results and theories within these areas, most educators agree, in principle if not always in practice, that it is more important that these courses help students understand how science is done, and when and why the deliverances of scientific inquiry should be trusted. Such understanding is essential not just to students who aspire to be scientists, but to all of us, since the claims of science inevitably shape our society and politics and touch on our public and private lives.

Just how we should characterize the nature of science, or indeed whether science has a nature at all, has been a matter of debate for fifty years or more (Bell et al., 2001). NOS has traditionally been understood to be concerned with the nature of scientific methods, but more broadly with the norms, values, attitudes, and institutions that shape scientific activity. To understand these things, science educators have often looked to philosophers, historians, and sociologists of science for guidance. Early formulations of NOS draw heavily on the work of logical empiricists and on Karl Popper's falsificationism. Later iterations of NOS sought to incorporate the insights of the historical turn in the history and philosophy of science, emphasizing, for instance, Kuhn's notion of scientific paradigms.

The essays in this volume are inspired by a more recent turn in the history of philosophy of science. Beginning in the 1990s, a group of philosophers, now often called new mechanists, observed that theories of the nature of science that had dominated twentieth century philosophy of science were strangely silent about the concept of mechanism (Craver et al., 2024). "Mechanism" is one of the most widely used terms in the scientific vocabulary and many scientists would describe the identification of mechanisms as a central aim of their research. Even so, philosophers had said precious little about what mechanisms are and how scientists identify and characterize them. The new mechanists set about rectifying this omission, developing mechanism-focused accounts of discovery, modeling, reasoning, evidence, and explanation (Glennan, 2017; Glennan et al., 2021). Collectively these efforts offer science educators a new image of NOS that speaks more directly to the practice of modern science.

To understand the centrality of mechanistic reasoning to the practice of science, it is useful to distinguish three tasks in scientific inquiry: first, describing and delineating phenomena; second, identifying causes of these phenomena; and third, characterizing the mechanisms responsible for these phenomena. None of these tasks are straightforward.

Consider global warming. Characterizing the phenomena involves identifying variables that measure changes in global temperature over time. This work requires complex theoretical and empirical methods to separate stable patterns of climate change from the exigencies of local weather, and to identify the historical traces that allow scientists to characterize this data over centuries or millennia. Once the phenomenon of global warming has been established, scientists look for events or factors that may have causally contributed to this change—for instance, fossil fuel consumption, volcanic eruptions, or variations in solar activity. Additionally, scientists try to identify the mechanisms by which these causal factors influence global warming—for instance, showing how fossil fuel consumption leads to greenhouse gas emissions that are trapped in the atmosphere and reduce the amount of heat escaping from the atmosphere.

Causes are things (properties, events, facts...) that make a difference to their effects. To say, for instance, that burning fossil fuels causes global warming is to say that this burning has made a difference to global temperatures. The difference-making criteria implies that causes are things we can manipulate. Had we burned fewer fossil fuels, the atmosphere would not have warmed so much, and reducing fossil fuel emissions in the future will slow the rate of climate change. But while knowledge of causes tells us *what* some effect depends on, it does not tell us *why* or *how* the cause influences the effect. For that, we need mechanisms. Mechanisms, the new mechanists tell us, are collections of entities whose activities and interactions produce, underlie, or sustain some phenomenon. They are the processes by which a cause brings about an effect.

When one thing causes another, there must be a mechanism by which this cause brings about an effect. Consider a second example—the effect of adding manure to soil. Simple experiments can establish that adding manure is a difference maker that improves plant growth, but these experiments leave untouched the question of why and how manure has that effect. To understand that, we must understand (among many other things) that manure is a source of nitrogen, that nitrogen is an essential constituent in chlorophyll, and that chlorophyll is an entity that plays a central role in the mechanism of photosynthesis.

The picture painted here suggests that scientists seek three kinds of knowledge—knowledge of happenings in some domain (phenomena), knowledge about causes that make a difference to these phenomena, and knowledge of the mechanisms responsible for this difference-making. While the process of discovery typically leads from phenomena to causes to mechanisms, the actual process can have many entry points and iterations. Sometimes we are able to characterize causes while knowing little or nothing about mechanisms, while in other cases we need mechanistic knowledge to establish causal connections. Sometimes awareness of possible mechanisms leads us to try to identify as yet unnoticed phenomena. Sometimes mechanistic knowledge leads us to recognize that we have characterized a phenomenon and its causes incorrectly. Sophisticated mechanistic reasoning is not necessarily linear.

Learning how to reason mechanistically will help students understand how the world is organized, and how individuals and societies can act well and wisely in it. They can recognize first and foremost the nested nature of the world's causal structures. Mechanisms are composites—made of entities and activities, and these components of mechanisms are themselves decomposable into further entities and activities. They can learn how timing and other forms of organization matter in nature, how the same components differently arranged produce radically different effects. And even as they appreciate the reductionist impulse in mechanistic reasoning, they can, as they progress, learn that what mechanisms do depend upon how they are situated, so that to understand the world's causal structure, scientists must not just look down, but also up and around.

Understanding how causes depend upon mechanisms can help students appreciate both the sources and limitations of general causal knowledge. Most of the generalizations we find in science—from generalizations about the products and rates of chemical reactions, to generalizations about cellular metabolism or gene flow in populations, to generalizations about stellar evolution—are really just descriptions of how mechanisms behave. Understanding this fact can help them to see that these generalizations hold only so long as the underlying mechanisms are not broken or interfered with. It can help them recognize, for instance, how runaway growth in cancer depends upon the breakdown of mechanisms regulating cell division, or how changes to climate can alter mechanisms that in ways that change formerly regular patterns of rainfall.

Mechanistic knowledge is the ground of informed intervention into the world. Knowing something about the mechanisms of transmission of a virus, for instance, is crucial to act to prevent its spread. Or for that matter, it is likely that understanding the mechanisms by which students learn is key to sensibly developing curricula and pedagogies that achieve our goals.

There is one last lesson about mechanisms that both we and our students would do well to learn: mechanisms are particular things, and no one is quite like the other. Scientists rightly prize general knowledge, and often characterize mechanisms in general ways. Neuroscientists, for instance, speak of *the* mechanism of the action potential. But the definite article can deceive us. As any neuroscientist will admit, there are countless varieties of neurons, and both the behavior and underlying mechanisms of action potentials vary across organisms, and across neurons within organisms. To understand mechanisms like these, we necessarily introduce abstraction and idealization into our models, but successful mechanistic reasoners will always remember that the mechanisms their models describe vary in countless ways not captured in their representations.

The chapters in this book combine theoretical and empirical work to provide a wealth of insights into how mechanistic reasoning works in a variety of different scientific domains, and how educators can facilitate and assess student learning about this key feature of scientific inquiry. Application of this research to the classroom promises to help science students at all levels think more clearly about mechanisms. Developing this ability can help them become more informed and effective producers and consumers of scientific knowledge.

Butler University, Indianapolis
IN, USA

Stuart Glennan

References

Bell, R., Abd-El-Khalick, F., Lederman, N. G., Mccomas, W. F. & Matthews, M. R. (2001). The nature of science and science education: A bibliography. *Science & Education, 10*(1), 187–204. https://doi.org/10.1023/a:1008712616090.

Craver, C., Tabery, J., & Illari, P. (2024). Mechanisms in science. In E. N. Zalta, U. Nodelman (Eds.) *The Stanford encyclopedia of philosophy* (Fall 2024). Metaphysics Research Lab, Stanford University. https://plato.stanford.edu/archives/fall2024/entries/science-mechanisms/.

Glennan, S. S. (2017). *The new mechanical philosophy*. Oxford University Press.

Glennan, S. S, Illari, P., & Weber, E. (2021). Six theses on mechanisms and mechanistic science. *Journal for General Philosophy of Science*, no. 0123456789.

Holbrook, J. (2010). Education through science as a motivational innovation for science education for all. *Science Education International, 21*(2), 80–91.

NSTA Board of Directors. (2020). Nature of Science | NSTA. https://www.nsta.org/nstas-official-positions/nature-science.

Preface

Due to their centrality in science and scientific information, mechanisms and mechanistic explanations have long been a focal point of interest within the science education community. However, the absence of a well-defined framework for describing mechanistic explanations and their components led to contradictions in characterizing the challenges of this practice and identifying effective ways to overcome them. The publication by Russ et al. (2008) addressed this gap by building on definitions of mechanisms from the philosophy of science. This pivotal work provided clarity and led to a shared foundation, sparking a surge of interest in the field of mechanistic reasoning.

Since then, numerous studies in science education have embraced this framework, uncovering the challenges that mechanistic reasoning poses for students, teachers, and even laypersons. These studies identified scaffolding strategies to address these challenges and demonstrated the importance of mechanistic reasoning in supporting decision-making, understanding the nature of science, and fostering broader scientific literacy.

Today, with a substantial body of knowledge on mechanistic reasoning, it is the right time to consolidate what we know about the topic in science education across different scientific disciplines, including biology, chemistry, and physics. This edited book delves into the realm of mechanistic reasoning within science education, exploring its significance, development, and practical applications across various studies in these scientific disciplines. Structured into four distinct parts, the book examines this theme through diverse lenses, including research studies, instructional strategies, the nature of explanations, and its overall impact on scientific literacy.

The first section on *Capturing and Evaluating Mechanistic Reasoning* addresses methods for capturing and evaluating mechanistic reasoning in educational settings. It begins with a chapter by Watts, Dood, and Shultz, which explores the use of AI models to automatically analyze students' writing assignments in organic chemistry. The following chapter by Trommler and Hammann examines the differences between mechanistic and non-mechanistic teleological explanations in biology. In this chapter, they demonstrate how domain-specific knowledge influences the construction of mechanistic and non-mechanistic teleological explanations. Martin, Yik, and Graulich examined how undergraduate organic chemistry students judge the plausibility of mechanisms in argumentation tasks, revealing that initial claims for or against a reaction product in students' argumentation significantly predict student mechanistic reasoning complexity. The final chapter in this section by de Andrade, Shwartz, and Rodrigues provides a more broader characterization of mechanistic reasoning by highlighting features that can guide mechanistic reasoning across chemistry, physics, and biology.

The second section on *Promoting Mechanistic Reasoning* explores approaches for promoting mechanistic reasoning from various perspectives. It presents a broad spectrum of pedagogical practices, including embodiment, creating animations, and identifying knowledge gaps. The first chapter begins with the study by Mathayas and Lindgren demonstrating that the human body, e.g., hand gestures, provides a valuable aid in facilitating mechanistic reasoning. The chapter by Cooper, Franovic, and Noyes discusses instructional practices that promote mechanistic reasoning in undergraduate science courses from a broader perspective, emphasizing the importance of scaffolded, evidence-centered formative assessments to elicit and enhance students' reasoning. Livni Alcasid and Haskel-Ittah describe the affordances and constraints of explicit teaching of black boxes, the gaps included in all mechanistic explanations. The last chapter in this section by Bachtiar, Meulenbroeks, and van Joolingen explores the use of stop-motion animations to support students in developing mechanistic reasoning, with a particular focus on the quality of stop-motion animations in relation to students' mechanistic reasoning.

The third section focuses on *Constructing Mechanistic Explanation* and the cognitive processes and affordances involved in constructing mechanistic explanations. The chapter by Tang sheds light on mechanistic reasoning in elementary science classrooms. In a case study, Tang examines how mechanistic reasoning supports explanation-building in fourth graders exploring sound generation and transmission. The analysis highlights constraints such as prematurely accepted explanations and limited reasoning resources. Talanquer explores the complexity of mechanistic reasoning in chemistry and the affordances of representations, such as multiple levels of granularity, which are used to express the causal story of a mechanism. The last chapter by Fuhrmann, Rosenbaum, Wagh, Eloy, Blikstein, and Wilkerson dives deep into an instructional sequence, which uses different modalities to support mechanistic reasoning. Thereby, they examined how sixth-grade students expressed elements of mechanistic reasoning during a computational modeling unit as a function of the modalities used, such as verbal or written assignments.

The final section, *Implications for Scientific Literacy,* explores the broader implications of mechanistic reasoning for the use of science in everyday life. The first chapter by Stowe, Russ, and Wells compares mechanistic reasoning in everyday life with the reasoning required by students in the science classroom. They raise the question of what is necessary to prepare students for their life after school and how to enable students to engage in reasoning that more closely resembles everyday life. In the second chapter, Dubovi demonstrates how mechanistic reasoning in everyday life is connected to health-related decision-making.

This book provides a comprehensive exploration of mechanistic reasoning in science education, addressing a significant gap in the literature despite growing interest in the topic. Combining theoretical discussions with empirical evidence and practical applications, it offers a multifaceted perspective for science educators, curriculum designers, and teachers. The book's practical recommendations for evaluating and promoting mechanistic reasoning aim to enhance teaching practices and contribute to related fields, such as explanations, systems thinking, modeling, and argumentation.

Department of Science Teaching
Weizmann Institute of Science
Rehovot, Israel
Michal Haskel-Ittah

Institute of Chemistry Education, Justus-Liebig-University
Giessen, Germany
Nicole Graulich

Faculty of Teacher Training and Education
University of Jember
Jember, Indonesia
Rayendra Bachtiar

Contents

Part I Capturing Reasoning About Mechanisms—Analyzing and Evaluating Mechanistic Reasoning

1 Using Unsupervised Machine Learning to Explore Organic Chemistry Students' Mechanistic Reasoning in Writing-to-Learn Assignments 3
Field M. Watts, Amber J. Dood, and Ginger V. Shultz

2 Students' Teleological Reasoning: An Explanatory Escape 21
Friederike Trommler and Marcus Hammann

3 Mechanistic Reasoning About Alternatives: How Students Reason About Plausible and Implausible Mechanisms in Chemistry . . . 41
Paul P. Martin, Brandon J. Yik, Benjamin T. Burlingham, and Nicole Graulich

4 Towards a Wide Understanding of Mechanisms Across Science Domains: Implications for Science Education 63
Vanessa De Andrade, Yael Shwartz, and Pedro Correia Rodrigues

Part II Promoting Mechanistic Reasoning—Implications for Supporting Mechanistic Reasoning

5 Facilitating Mechanistic Reasoning with Gesture-Augmented Simulations: A Study of Two Students' Sensemaking of Air Pressure 85
Nitasha Mathayas

6 Design Principles for Developing and Analyzing Formative Assessments to Support Mechanistic Reasoning 107
Keenan Noyes, Clare G. -C. Franovic, and Melanie M. Cooper

7 Explicating Black Boxes in the Biology Classroom for Promoting Mechanistic Reasoning 125
Gur Arie Livni Alcasid and Michal Haskel-Ittah

8 Promoting Students' Mechanistic Reasoning in Physics Through Constructing Stop-Motion Animations: What to Consider? 145
Rayendra Wahyu Bachtiar, Ralph F. G. Meulenbroeks, and Wouter R. van Joolingen

Part III Processes of Constructing Mechanistic Explanations

9 On the Productivity of Mechanistic Reasoning in Elementary Science Classroom 161
Xiaowei Tang, Deying Wen, and Ning Pan

10 The Nature of Mechanistic Reasoning in Chemistry 179
Vicente Talanquer

11 Modality Matters: Exploring and Capturing Students' Mechanistic Reasoning During Computational Modeling in Science Education 197
Tamar Fuhrmann, Leah F. Rosenbaum, Aditi Wagh, Adelmo Eloy, Paulo Blikstein, and Michelle H. Wilkerson

Part IV Contribution of Mechanistic Reasoning to Scientific Literacy

12 Mechanistic Reasoning In-School Versus Mechanistic Reasoning In-Life 223
Ryan L. Stowe, Rosemary S. Russ, and Lindsay B. Wells

13 Mechanistic Literacy Across Functional Health-Related Daily Life Decisions 239
Ilana Dubovi

Part I
Capturing Reasoning About Mechanisms—Analyzing and Evaluating Mechanistic Reasoning

Chapter 1
Using Unsupervised Machine Learning to Explore Organic Chemistry Students' Mechanistic Reasoning in Writing-to-Learn Assignments

Field M. Watts, Amber J. Dood, and Ginger V. Shultz

1.1 Introduction

Developing skills for mechanistic reasoning is a common learning outcome across STEM courses, and it is important for instructors to elicit and assess mechanistic reasoning to support students in achieving this learning goal (Bachtiar et al., 2022). Writing-to-learn (WTL) is one pedagogical approach which has been shown to promote students' conceptual learning and disciplinary thinking across STEM courses (Finkenstaedt-Quinn et al., 2023; Gere et al., 2019), including eliciting students' mechanistic reasoning in introductory organic chemistry (Finkenstaedt-Quinn et al., 2024; Watts et al., 2020, 2022a, b, 2024; Zaimi et al., 2024). For open-ended assessments such as WTL, recent studies have investigated using machine learning (ML) to assess student responses within STEM courses (Zhai et al., 2020). This includes several studies utilizing ML to investigate students' mechanistic reasoning in chemistry (Martin & Graulich, 2023). These studies use supervised ML to score short responses or unsupervised ML to analyze patterns in students' reasoning (Martin et al., 2023; Martin & Graulich, 2023; Zhai et al., 2020). The goal of the present study is to leverage a pipeline of supervised and unsupervised ML analysis to afford insights regarding the nature of students' mechanistic reasoning in the context of an organic chemistry WTL assignment.

F. M. Watts (✉)
Department of Chemistry & Biochemistry, University of Wisconsin-Milwaukee, Milwaukee, WI, USA

A. J. Dood · G. V. Shultz
Department of Chemistry, University of Michigan, Ann Arbor, MI, USA

N. Graulich et al. (eds.), *Exploring Mechanistic Reasoning in Science Education*, Contributions from Science Education Research 17,
https://doi.org/10.1007/978-3-032-10246-1_1

1.1.1 Eliciting Mechanistic Reasoning in Organic Chemistry

Organic chemists use mechanistic reasoning to predict and explain reactions between compounds (Goodwin, 2003). As such, supporting students' mechanistic reasoning represents a key learning goal which aligns with the scientific practices of constructing explanations and reasoning about models (National Research Council, 2012). Nevertheless, mechanisms remain one of the more challenging topics for organic chemistry students (Dood & Watts, 2022, 2023). Across organic chemistry educational studies, researchers have used various frameworks to investigate students' mechanistic reasoning (Dood & Watts, 2022). Commonly-used frameworks build upon work within the philosophy of science regarding what a mechanism entails (Machamer et al., 2000; Russ et al., 2008). These frameworks establish mechanistic reasoning as explaining a phenomenon by attending to the changes that occur within the system at a scalar level below the phenomenon of interest (Machamer et al., 2000; Russ et al., 2008). In the context of organic chemistry reactions, mechanistic reasoning entails describing the interactions between molecules, atoms, and electrons which result in a particular reaction product. Studies investigating organic chemistry students' mechanistic reasoning used these frameworks in the context of several tasks (Caspari et al., 2018; Keiner & Graulich, 2020, 2021), including WTL assignments that elicit students' explanations of reaction mechanisms (Watts et al., 2020, 2024).

Within the context of WTL, research demonstrates that assignments with peer review and revision elicit students' writing about specific features necessary for engaging in mechanistic reasoning (Finkenstaedt-Quinn et al., 2024; Watts et al., 2020, 2022a, b, 2023b, 2024). Identifying the features of students' writing can allow for understanding how students reason about reaction mechanisms, as demonstrated in prior studies focused on an organic chemistry WTL assignment which elicited students' reasoning for a hydrolysis reaction yielding two distinct products (Watts et al., 2020, 2022a, 2024). In one study, students largely explained the formation of two products by indicating a divergence in the reaction pathway at the first step of the reaction rather than a more likely step later in the reaction (Watts et al., 2020). In another study, students' responses included reasoning based on both surface features (e.g., formal charges) and implicit properties (e.g., acidity/basicity), which was influenced by both assignment design and the peer review process (Watts et al., 2024). These findings resonate with the literature on students' mechanistic reasoning in organic chemistry, which indicates the importance of task design and the need to support students with building connections between the changes occurring during a reaction and the underlying chemical properties (Crowder et al., 2024; DeCocq & Bhattacharyya, 2019; Farheen et al., 2024; Houchlei et al., 2021). Broadly speaking, the literature suggests the need to elicit students' reasoning about reaction mechanisms by asking students *how* and *why* mechanisms occur, beyond asking students to describe *what* happens during a reaction (Dood & Watts, 2022).

1.1.2 Machine Learning in STEM Education

With calls for open-ended assessments that encourage students to provide explanations for scientific phenomena, researchers have increased focus on applying ML within STEM education (Zhai et al., 2020). Because ML allows for automated analysis of responses, ML represents an opportunity for addressing the barriers to implementing open-ended assessments (such as time constraints preventing instructors' ability to evaluate responses and provide feedback; Moon et al., 2018). Several studies explore using ML to investigate mechanistic reasoning (Martin & Graulich, 2023). In chemistry education, studies use supervised ML (which requires human-annotated data) to score or label student responses (Dood et al., 2018, 2020a, 2024; Martin & Graulich, 2024; Noyes et al., 2020; Watts et al., 2022a; Yik et al., 2021). At least one study leveraged unsupervised ML, which does not require human annotation, to investigate student responses (Martin et al., 2023). These ML models can be useful for instruction by categorizing student responses to provide adaptive interventions or immediate, personalized feedback (Dood et al., 2019, 2020b; Watts et al., 2023a; Yik et al., 2023).

In this article, we present the use of unsupervised ML to analyze student responses to organic chemistry WTL assignments eliciting students' mechanistic reasoning. This study extends our prior work which developed supervised ML models to identify specific mechanistic reasoning features, such as descriptions of charges or electron movement, within students' responses to the aforementioned WTL assignments (Watts et al., 2022a). The approach to apply unsupervised ML analysis to the output of the supervised ML models was driven by the desire to address the limitation of the supervised ML models, which provide an analytic (rather than holistic) understanding of students' responses. While unsupervised ML could be applied directly to the raw text data, the approach used in the present study offers the benefit of deriving the unsupervised analysis from ML models based on existing frameworks, thereby overcoming a potential limitation of unsupervised ML which can lead to categorizations that may be difficult for a human to interpret (Zhai et al., 2020). The unsupervised ML approach identifies holistic characterizations of student responses based on identifiable features of mechanistic reasoning within their writing, which can allow instructors and researchers to develop individualized approaches to support students' learning.

1.2 Theoretical Perspectives

This study is guided by the cognitive process theory of writing, which conceptualizes writing as a process that involves recursively planning, drafting, and reviewing text (Hayes, 1996). The writing process is guided by the task environment, which includes the text-in-production and the rhetorical problem (e.g., topic and audience) addressed by the writing task. During the writing process, the writer accesses their long-term memory and any external resources (e.g., textbooks) to determine how to represent their knowledge. The cognitive process theory guided the design and implementation for the WTL assignments described in this study while also providing a basis for using student writing as evidence of their mechanistic reasoning. Specifically, the recursive writing processes are supported by required peer review and revision components of the WTL assignments. Through recursively accessing their knowledge to respond to the WTL assignment, students must choose specific features to incorporate into their writing to demonstrate their knowledge.

This work operationalizes mechanistic reasoning for the organic chemistry context using Russ et al.'s (2008) framework developed from the philosophy of science. In this framework, mechanistic explanations must account for the underlying *entities* (i.e., molecules, atoms, and electrons), the *activities* between entities (i.e., the movement of electrons which result in changes in bonding), and the *properties of entities* (i.e., chemical properties which explain why entities interact). Mechanistic reasoning also includes *chaining*, which involves connecting the steps of a mechanism based on what happened in previous steps or what will happen in future steps (Russ et al., 2008).

1.3 Research Questions

In our prior work, we developed supervised ML models to identify mechanistic reasoning features within students' WTL responses at the sentence level. These models can be used to identify the total number of sentences for each feature that each student included in their response. This analysis can be useful for providing students with automated, formative feedback about specific mechanistic reasoning features (Watts et al., 2023a). However, the focus of the analysis on individual features makes it difficult to provide more holistic feedback regarding students' reasoning. Hence, the present study seeks to leverage unsupervised ML to investigate the nature of student responses more holistically. Specifically, we describe using latent class analysis (LCA) to group student responses based on the distribution of mechanistic reasoning features within individual responses. We conducted LCA because the output offers a statistically robust model for determining the probabilities of class membership, which represents an advantage over similar unsupervised clustering approaches (e.g., hierarchical or k-means clustering; Weller et al., 2020). Ultimately, this research seeks to explore how unsupervised ML can extend the utility of supervised ML. In doing so, we provide a more holistic depiction of the range of students' mechanistic reasoning as elicited by organic chemistry WTL assignments. The following research questions guided this study:

RQ1. To what extent can unsupervised ML be leveraged to categorize student responses to WTL assignments based on the features of mechanistic reasoning identified in their responses?

RQ2. How do the ML categorizations provide insights regarding the attributes of students' mechanistic reasoning?

1.4 Methods

1.4.1 Setting and Participants

The study took place within a second-semester, introductory organic chemistry laboratory course at a large, research-intensive university in the Midwestern United States. The study involved data collected from four semesters of the course (2018, 2019, 2020, and 2022). The study received Institutional Review Board approval, and participants were students who provided consent for data collection as part of a larger research effort to investigate the WTL program. The laboratory is separate from the second-semester lecture course, and students typically enroll in both concurrently. The laboratory course includes a weekly, one-hour lecture that covers topics related to the weekly, four-hour laboratory sessions. Topics covered include specific reactions, interpreting spectra, and chemical reasoning. The course is primarily taken by non-chemistry majors, typically in students' first or second years. Across all four semesters, the WTL assignments were administered asynchronously through the online learning management system.

1.4.2 Writing-to-Learn Assignment and Implementation

The WTL assignment design and administration followed the structure provided by a broader program implementing WTL across introductory STEM courses (Finkenstaedt-Quinn et al., 2021). The assignment varied each semester, reflecting an iterative assignment development process; the variations are detailed in the Supporting Information. All iterations were guided by the effective components of WTL assignments by including a meaning-making task and clear expectations (Gere et al., 2019). The meaning-making task was based on the history of the thalidomide molecule, which was originally used to treat morning sickness until it was discovered that racemization produces a stereoisomer with negative side-effects. The goal for the assignment was for students to explain two reactions that affect the thalidomide molecule (racemization and hydrolysis) and to design an analogous compound that would prevent these reactions. Students were provided a figure showing thalidomide and the two hydrolysis products (Fig. 1.1). In the most recent iterations of the assignment (2020, 2022-A, and 2022-B), the language eliciting the explanation of the two mechanisms stated,

> "Provide thorough descriptions of the mechanisms of both racemization and hydrolysis, highlighting the critical structural features of Thalidomide and their role in these mechanisms. When racemization occurs, what changes occur in the molecule? When hydrolysis occurs, what changes occur in the molecule?"

The expectations for the assignment were included through a set of rubric criteria provided to students and within the assignment itself as a list of what students should include.

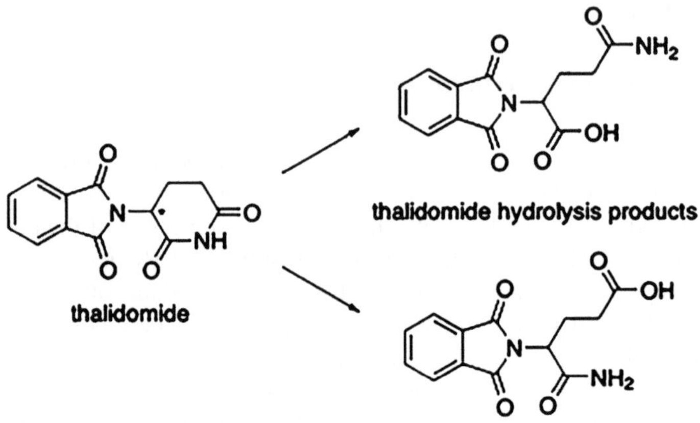

Fig. 1.1 Thalidomide and two hydrolysis products

The WTL assignments were administered similarly across all semesters. Assignment implementation adhered to two additional components of effective WTL assignments: providing opportunities for interactive writing and promoting metacognition (Gere et al., 2019). These were achieved by requiring students to submit a first draft, undergo peer review, and submit a revised draft. During peer review, students provided feedback to three peers by responding to content-focused questions aligned with the assignment expectations. Through peer review, students engaged in interactive writing by receiving feedback and responding to their peers' writing. The required revision promoted students' metacognition by encouraging them to reflect on their writing and the decisions they made when constructing their responses.

1.4.3 Data Collection

The data collected includes the N = 4740 responses to all assignment iterations, including initial and final drafts. Sample sizes for each iteration are provided in the Supporting Information.

1.4.4 Data Analysis

To prepare the dataset for LCA, students' responses were analyzed through a pipeline including supervised ML models (Fig. 1.2). The first step used ML to identify whether sentences were relevant to their explanations of the reaction mechanisms. This model, a convolutional neural network, was developed in a prior study and achieved 88.6% accuracy and 0.762 Cohen's κ, exceeding the recommended values for using ML in assessment (Watts et al., 2024; Williamson et al., 2012). Next, relevant sentences were analyzed by a set of models which identified various features necessary for mechanistic reasoning.

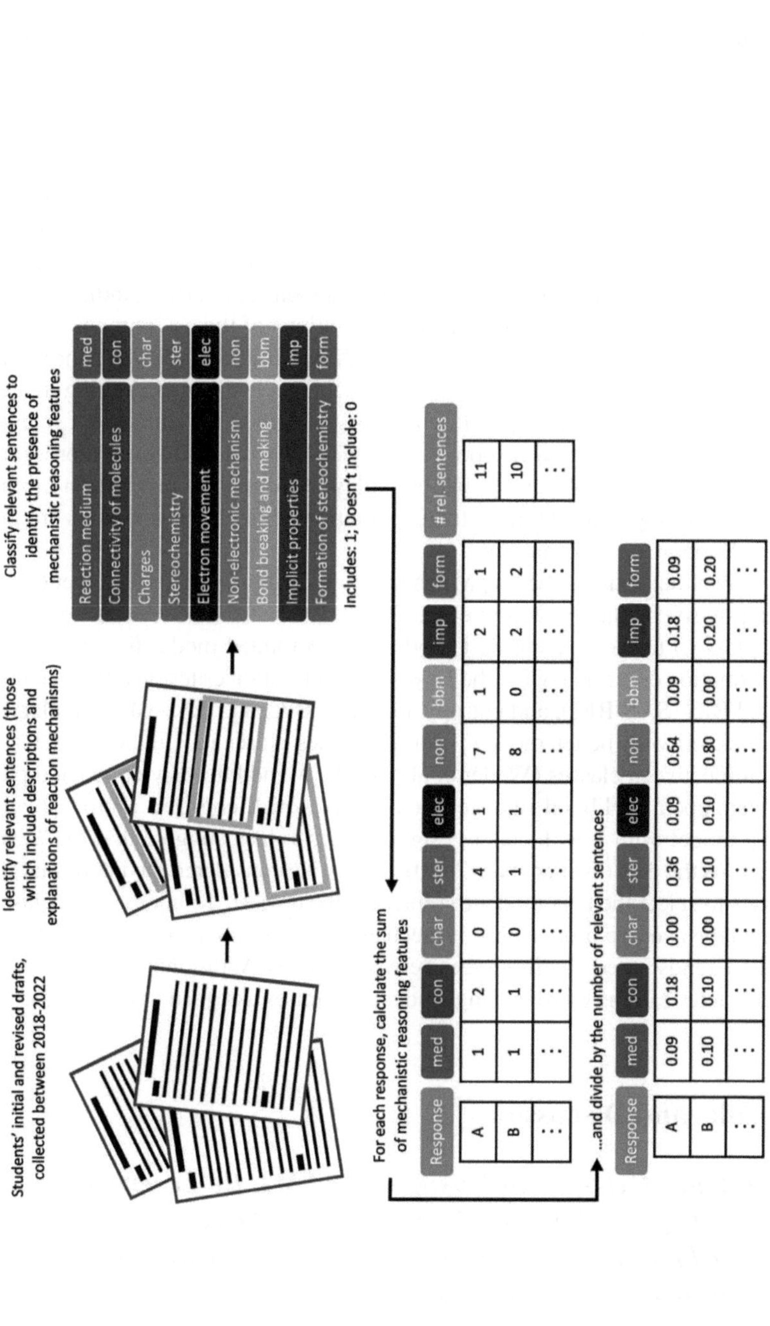

Fig. 1.2 Pipeline to prepare students' writing for LCA, resulting in a dataset that represents each response with the proportions of mechanistic reasoning features present

The models for identifying features of mechanistic reasoning were developed using students' final drafts in response to the *2020* assignment, as detailed in an earlier report (Watts et al., 2022a). These models align with Russ et al.'s (2008) framework and allow for identifying mechanistic reasoning features at the sentence level. Details regarding the application of these models for the present study (including establishing human-machine agreement for the assignment iterations not included during the original model development) are provided in the Supporting Information. In general, the agreement measures approach or exceed the recommended values for using ML in assessment (Williamson et al., 2012), indicating that the models can be applied to the data from all iterations of the assignment.

For the final step to prepare the data for LCA, we transformed each individual student response by calculating the sum of each mechanistic reasoning feature (across the single response) and dividing each sum by the total number of relevant sentences within the individual response. This yielded a dataset which represented each response as a vector of continuous variables reflecting the proportion of each mechanistic reasoning feature within the explanation. Descriptive statistics are provided in the Supporting Information.

LCA was conducted using the StepMix package in Python, which implements bias-adjusted step-wise estimators (Morin et al., 2024). We tested six LCA models (from one to six classes) to identify the best model. We examined model fit based on the Akaike information criterion (AIC), the Bayesian information criterion (BIC), sample-size adjusted BIC (SAS-BIC), and entropy. For AIC, BIC, and SAS-BIC, smaller values are associated with better models; entropy values closer to 1 indicate increased differentiation between classes (Weller et al., 2020). We additionally conducted parametric bootstrapped likelihood ratio tests (BLRTs), which compare a model with k classes to a model with $k - 1$ classes; significant p-values suggest that the k-class model better fits the data (Weller et al., 2020). Lastly, we considered the interpretability of the classes to select the best model to represent the dataset. To interpret the classes, we examined the average proportions of mechanistic reasoning features for each class and identified trends based on differences between classes. Additionally, we confirmed the class characterizations by reviewing student responses assigned to each class.

1.5 Results and Discussion

1.5.1 To What Extent Can Unsupervised ML Be Leveraged to Categorize Student Responses to WTL Assignments Based on the Features of Mechanistic Reasoning Identified in Their Responses?

Fit criteria for the LCA models with increasing numbers of classes are presented in Table 1.1. Recommendations for identifying the best model suggest selecting that which minimizes the BIC (Weller et al., 2020); as such, the six-class model appears

Table 1.1 Fit criteria for the LCA solutions

Number of classes	LL	AIC	BIC	SSA-BIC	Frequency of smallest class (%)	Entropy	BLRT p-value
1	27079.88	−54123.76	−54007.42	−53989.42	–	–	–
2	28976.61	−57879.22	−57640.06	−57444.47	0.21	0.8046	0.00
3	30114.79	−60117.57	−59755.60	−59459.58	0.18	0.7168	0.00
4	32513.86	−64877.73	−64392.94	−63996.48	0.14	0.7291	0.00
5	37094.55	−74001.09	−73393.50	−72896.6	0.12	0.8414	0.00
6	37353.00	−74480.00	−73749.59	−73152.26	0.08	0.8358	0.09

LL log-likelihood, *AIC* Akaike information criterion, *BIC* Bayesian information criterion, *SSA-BIC* sample-size adjusted BIC, *BLRT* parametric bootstrapped likelihood ratio rest

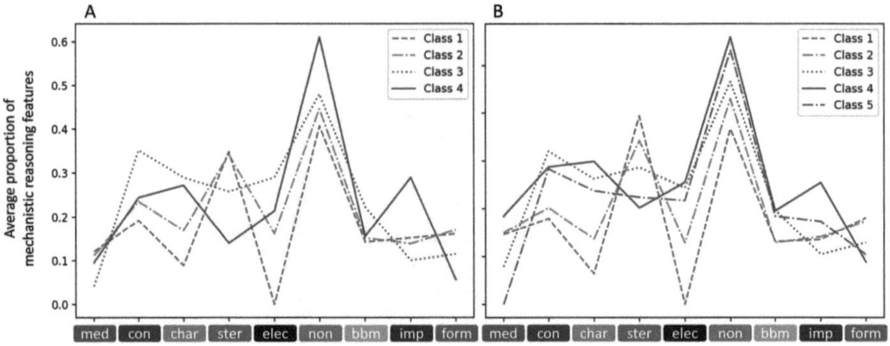

Fig. 1.3 Average proportions of each mechanistic reasoning feature across the classes, for the (**a**) four-class and (**b**) five-class LCA models. Definitions for the abbreviations along the x-axis are shown in Fig. 1.2

to be the best solution. However, the BLRTs indicate that increasing the number of classes from five to six does not provide a better model (due to the non-significant p-value for the six-class solution).

Given the BLRT, we examined the average proportion of mechanistic reasoning features for each class to determine the interpretability of the five-class solution. For parsimony, we additionally compared the five-class and four-class solutions. Based on the degree of distinction between separate classes, the four-class solution afforded the most interpretability (Fig. 1.3).

1.5.2 How Do the ML Categorizations Provide Insights Regarding the Attributes of Students' Mechanistic Reasoning?

We characterized each class by examining the overall pattern of the average proportions of mechanistic reasoning features and the key differences between other classes (Fig. 1.3a). Additionally, we reviewed responses assigned to each class to consider qualitative differences. Class characterizations are provided in Table 1.2, which includes the class sizes and exemplar responses to demonstrate the alignment between the interpretations and the qualitative differences in student responses.

The four classes suggest a hierarchy of sophistication for students' descriptions of mechanisms, from responses with minimal inclusion of electron movement (class one), a key component of organic chemistry mechanisms (Bhattacharyya, 2013; Duis, 2011); to responses which describe *how* the mechanism occurs by accounting for electron movement (class two); to responses which additionally include *why* the mechanism occurs by either emphasizing surface features (class three) or implicit features (class four). The classes relate to the framework for characterizing how students engage in mechanistic reasoning (Russ et al., 2008) by differentiating

Table 1.2 Characterization of the four classes, including the number of responses within each class, the name given to capture the quality of responses in each class, a brief interpretation of the key characteristics which differentiate classes from one another, and an exemplar response from each class

Class	Number of Responses (%)	Class name	Class interpretation	Exemplar response (emphasis added)
1	695 (14.7%)	Little mechanistic description	Responses emphasize stereochemistry, with the least emphasis on features more pertinent to describing the reaction mechanism (electron movement, connectivity, and charges).	"Our first concern is racemization, which is the inversion of a **stereocenter** within this molecule that causes the birth defects. Racemization is when there is rotation around one of the bonds within the molecule that has 4 different atoms connected to it creating an **enantiomer**. In this case the stereocenter rotating is the carbon-nitrogen bond where you can see a wedge and a dash... The mechanism is shown above, however I will outline the main steps as well. In general terms, the **stereocenter carbon** gets deprotonated, and after some charge stabilization/resonance structures, it gets re-protonated, but from the opposite side of the molecule resulting in **inversion of stereochemistry**. Finally protonation of this new double bond occurs opposite to the side that originally has the carbon, resulting in the **unfavorable enantiomer**."
2	898 (18.9%)	Emerging mechanistic description	Responses demonstrate a similar pattern of features compared to class 1, with inclusion of more features relevant to describing the reaction mechanism (electron movement, connectivity, and charges).	"Racemization occurs at a neutral pH creating equal proportions of the two enantiomers. The racemization occurs when water becomes an H_3O^+ molecule through protonation and protonates the **carbonyl functional group next to the stereocenter on the leftmost ring**. The water molecule can then deprotonate the hydrogen at the chiral stereocenter **resulting in the electrons forming the hydrogen-carbon bond** to create a double bond between the carbon that was originally a stereocenter and the **carbon connected to the protonated carbonyl** and then force the **electrons** from the double-bond constructing the carbonyl up on to the oxygen atom to eliminate its **positive charge**. When the H_3O^+ molecule re-protonates what used to be a chiral stereocenter, the hydrogen now has a choice depending on its location in the solution. It can either re-protonate from the front face or the back face... A water molecule then deprotonates the oxygen atom and the **hydrogen's electrons reproduce the carbonyl functional group**. The possibility of adding to either face of the molecule creates equal proportions of both enantiomers in the product, which is known as a racemic mixture."
3	1549 (32.7%)	Mechanistic description—focus on surface features	Responses demonstrate mechanistic descriptions by including discussion of election movement, while emphasizing descriptions of connectivity and charges (surface features of reaction mechanisms) and placing less emphasis on implicit properties.	"In order to prevent the teratogen from being formed through racemization, we must discuss the mechanism by which such enantiomer is formed. At a neutral pH, the **oxygen double bonded to the carbon** directly right of the stereocenter attacks an H_3O^+ molecule, making an O-H bond and gaining a **positive charge**. Then, the **hydrogen attached to the stereocenter** is attacked by H_2O, breaking the C-H bond and **pushing the electrons** from the bond to form a double bond with the adjacent carbon, bonded with oxygen, **pushing the electrons** from its double bond with oxygen onto the oxygen, **thereby neutralizing it**. Then, the carbon-carbon double bond attacks an H_3O^+ molecule, creating a carbon-hydrogen bond from the top face of the molecule, and a **pair of electrons** from the OH are pushed back into a carbon-oxygen double bond. The OH is deprotonated, meaning it loses an electron, by water. The result is the enantiomer of the molecule."

(continued)

Table 1.2 (continued)

Class	Number of Responses (%)	Class name	Class interpretation	Exemplar response (emphasis added)
4	1598 (33.7%)	Mechanistic description—focus on implicit properties	Responses demonstrate mechanistic descriptions by including discussion of election movement, while emphasizing non-electronic mechanistic descriptions (e.g., using words including "attacks") and implicit properties (e.g., acidity/basicity, nucleophilicity/electrophilicity, resonance).	"The carbonyl groups are the double-bonded oxygens found attached to the ring… because the oxygen molecule is **electron deficient**, the proton adjacent to it is affected and will be more susceptible to deprotonation as it is more **acidic**. The second step is to take a water (H_2O) molecule and attack the **acidic proton** to deprotonate. Now you have **electrons that will form a double bond** between the two carbons and the double-bonded electrons will move up to the oxygen atom to create a **stabilized resonance form**. Since the carbon-hydrogen bond shifts to form a double bond in between the two carbon atoms, **resonance** pushes the carbon-oxygen double bond into a lone pair of electrons on the oxygen atom which is important for the next step. Next, the **delocalized electrons** on the oxygen atom will move up to recreate the carbonyl double bond prompting the carbon-carbon double bond to deprotonate the hydronium ion. The hydrogen can come from either direction which gives us our 50/50 racemization mixture."

between responses based on (1) the degree to which the response identifies electronic activities of entities (prominent in classes two, three, and four) and (2) the properties of entities used to explain the reaction (with class three focusing on surface-feature properties and class four focusing on implicit properties). Furthermore, the classes align with different levels of engagement with mechanistic reasoning reported in the organic chemistry education literature (Cooper et al., 2016; Graulich et al., 2019; Sevian & Talanquer, 2014), ranging from more descriptive responses (class one) to causal mechanistic responses which include more chemistry-based explanations for *why* the mechanisms occur by attending to implicit properties (class four). In addition to contributing to the literature by demonstrating how ML can be used to meaningfully interpret students' mechanistic reasoning across large datasets of open-ended responses, the four classes reflect prior studies which suggest that organic chemistry students are capable of reasoning with implicit properties but often focus on the surface features of reaction mechanisms (Anzovino & Bretz, 2016; Haas et al., 2024).

1.6 Limitations

There are several limitations associated with this research. First, the findings are limited to the specific WTL prompt analyzed; the LCA model and class characterizations may not apply to other assignments. Additionally, data were collected from a single institution and the findings may not capture alternative approaches for explaining reaction mechanisms. Furthermore, the approach for characterizing mechanistic reasoning may not fully capture all aspects of students' mechanistic reasoning; as described in our prior work (Watts et al., 2020, 2024), the *chaining* component of the mechanistic reasoning framework (Russ et al., 2008) was not elicited due to the nature of the WTL implementation. Lastly, while the LCA model provides useful insight regarding students' reasoning across a large corpus of WTL responses, we note that the quantitative methods may not reflect individual students' reasoning.

1.7 Implications

This study demonstrates that LCA can be used to categorize WTL responses based on the patterns of mechanistic reasoning features present in student responses. The categorizations from the LCA analysis provide holistic insight regarding the nature of students' reasoning. Knowing that students can engage in more sophisticated reasoning in organic chemistry, these findings provide several implications for instruction and future research to support students' developing mechanistic reasoning. These implications extend from the fact that the automated analysis pipeline and LCA model can be used to predict the class to which any given response belongs. This can allow for automated analysis of student responses with an output that provides both analytical detail about the mechanistic reasoning features present (based on the supervised ML models) and a holistic characterization of the response (based on the LCA categorization).

The automated analysis pipeline presented in this work can be useful for providing both instructor-facing and student-facing feedback. For example, instructor-facing feedback could take the form of a dashboard displaying the percentage of responses within each class. The dashboard would thereby provide the instructor insight regarding overall course performance, with the possibility for interactive features allowing instructors to read individual responses or to see how the ML models analyzed specific aspects of students' writing. Instructors could use such a dashboard to adjust instruction based on how students within their class responded to the assignment. Student-facing feedback could take the form of providing directive feedback based on the class to which a students' response belongs. For example, students with responses categorized as class two (emerging mechanistic description) could be directed to consider revising their response to emphasize features such as electron movement and implicit properties.

The possibilities for instructional implications can also lead to future research directions. For example, future studies could investigate how changes to the WTL assignment or implementation may influence students' mechanistic reasoning (e.g., analysis of students' writing across different prompts). Studies could also investigate how instructor- or student-facing feedback provided by the automated analysis pipeline can be used to support students with learning through the WTL process. Ultimately, the findings suggest the possibility for tools which leverage different types of ML to investigate and assess student writing in response to assignments eliciting mechanistic reasoning. Furthermore, while the present study focuses on mechanistic reasoning in the organic chemistry context, the analytical approach can be extended to broader science education contexts, from mechanistic reasoning in other disciplines to characterizing other forms of student reasoning (such as modeling or argumentation).

1.8 Conclusion

This work demonstrates the combination of supervised and unsupervised ML to characterize and interpret a large dataset of student responses to WTL assignments eliciting mechanistic reasoning. Building on prior work establishing supervised ML models that can reliably identify specific features of students' mechanistic reasoning at the sentence level (Watts et al.,, 2022a), our findings indicate how unsupervised ML (specifically LCA) can be used to holistically characterize student responses based on the distribution of mechanistic reasoning features present. These findings extend the literature focused on using ML to characterize mechanistic reasoning by highlighting how multiple approaches can be used together to investigate and characterize students' reasoning. The LCA model described in this study shows promise for being able to characterize student responses in such a way that can automatically provide instructors with an overview of how their students respond to the given WTL assignment. Furthermore, the automatic characterizations could prove useful for providing adaptive interventions or individual student feedback which could further support students' mechanistic reasoning.

References

Anzovino, M. E., & Bretz, S. L. (2016). Organic chemistry students' fragmented ideas about the structure and function of nucleophiles and electrophiles: A concept map analysis. *Chemistry Education Research and Practice, 17*(4), 1019–1029. https://doi.org/10.1039/C6RP00111D

Bachtiar, R. W., Meulenbroeks, R. F. G., & Van Joolingen, W. R. (2022). Mechanistic reasoning in science education: A literature review. *Eurasia Journal of Mathematics, Science and Technology Education, 18*(11), em2178. https://doi.org/10.29333/ejmste/12512

Bhattacharyya, G. (2013). From source to sink: Mechanistic reasoning using the electron-pushing formalism. *Journal of Chemical Education, 90*(10), 1282–1289. https://doi.org/10.1021/ed300765k

Caspari, I., Kranz, D., & Graulich, N. (2018). Resolving the complexity of organic chemistry students' reasoning through the lens of a mechanistic framework. *Chemistry Education Research and Practice, 19*(4), 1117–1141. https://doi.org/10.1039/C8RP00131F

Cooper, M. M., Kouyoumdjian, H., & Underwood, S. M. (2016). Investigating Students' reasoning about acid–base reactions. *Journal of Chemical Education, 93*(10), 1703–1712. https://doi.org/10.1021/acs.jchemed.6b00417

Crowder, C. J., Yik, B. J., Frost, S. J. H., Cruz-Ramírez De Arellano, D., & Raker, J. R. (2024). Impact of prompt cueing on level of explanation sophistication for organic reaction mechanisms. *Journal of Chemical Education, 101*(2), 398–410. https://doi.org/10.1021/acs.jchemed.3c00710

DeCocq, V., & Bhattacharyya, G. (2019). TMI (too much information)! Effects of given information on organic chemistry students' approaches to solving mechanism tasks. *Chemistry Education Research and Practice, 20*(1), 213–228. https://doi.org/10.1039/C8RP00214B

Dood, A. J., & Watts, F. M. (2022). Mechanistic reasoning in organic chemistry: A scoping review of how students describe and explain mechanisms in the chemistry education research literature. *Journal of Chemical Education, 99*(8), 2864–2876. https://doi.org/10.1021/acs.jchemed.2c00313

Dood, A. J., & Watts, F. M. (2023). Students' strategies, struggles, and successes with mechanism problem solving in organic chemistry: A scoping review of the research literature. *Journal of Chemical Education, 100*(1), 53–68. https://doi.org/10.1021/acs.jchemed.2c00572

Dood, A. J., Fields, K. B., & Raker, J. R. (2018). Using lexical analysis to predict Lewis acid–base model use in responses to an acid–base proton-transfer reaction. *Journal of Chemical Education, 95*(8), 1267–1275. https://doi.org/10.1021/acs.jchemed.8b00177

Dood, A. J., Fields, K. B., Cruz-Ramírez De Arellano, D., & Raker, J. R. (2019). Development and evaluation of a Lewis acid–base tutorial for use in postsecondary organic chemistry courses. *Canadian Journal of Chemistry, 97*(10), 711–721. https://doi.org/10.1139/cjc-2018-0479

Dood, A. J., Dood, J. C., Cruz-Ramírez De Arellano, D., Fields, K. B., & Raker, J. R. (2020a). Analyzing explanations of substitution reactions using lexical analysis and logistic regression techniques. *Chemistry Education Research and Practice, 21*(1), 267–286. https://doi.org/10.1039/C9RP00148D

Dood, A. J., Dood, J. C., Cruz-Ramírez De Arellano, D., Fields, K. B., & Raker, J. R. (2020b). Using the research literature to develop an adaptive intervention to improve student explanations of an S_N1 reaction mechanism. *Journal of Chemical Education, 97*(10), 3551–3562. https://doi.org/10.1021/acs.jchemed.0c00569

Dood, A. J., Watts, F. M., Connor, M. C., & Shultz, G. V. (2024). Automated text analysis of organic chemistry students' written hypotheses. *Journal of Chemical Education, 101*(3), 807–818. https://doi.org/10.1021/acs.jchemed.3c00757

Duis, J. M. (2011). Organic chemistry educators' perspectives on fundamental concepts and misconceptions: An exploratory study. *Journal of Chemical Education, 88*(3), 346–350. https://doi.org/10.1021/ed1007266

Farheen, A., Martin, N., & Lewis, S. E. (2024). Student perceptions of partial charges and nucleophilicity/electrophilicity when provided with either a bond-line, ball-and-stick, or electrostatic potential map for molecular representation. *Chemistry Education Research and Practice, 25*(1), 343–359. https://doi.org/10.1039/D3RP00173C

Finkenstaedt-Quinn, S. A., Petterson, M., Gere, A., & Shultz, G. (2021). Praxis of writing-to-learn: A model for the design and propagation of writing-to-learn in STEM. *Journal of Chemical Education, 98*(5), 1548–1555. https://doi.org/10.1021/acs.jchemed.0c01482

Finkenstaedt-Quinn, S. A., Watts, F. M., Shultz, G. V., & Gere, A. R. (2023). A portrait of MWrite as a research program: A review of research on writing-to-learn in STEM through the MWrite program. *International Journal for the Scholarship of Teaching and Learning, 17*(1), 1–15. https://doi.org/10.20429/ijsotl.2023.17118

Finkenstaedt-Quinn, S. A., Watts, F. M., & Shultz, G. V. (2024). Reading, receiving, revising: A case study on the relationship between peer review and revision in writing-to-learn. *Assessing Writing, 59*, 100808. https://doi.org/10.1016/j.asw.2024.100808

Gere, A. R., Limlamai, N., Wilson, E., MacDougall Saylor, K., & Pugh, R. (2019). Writing and conceptual learning in science: An analysis of assignments. *Written Communication, 36*(1), 99–135. https://doi.org/10.1177/0741088318804820

Goodwin, W. (2003). Explanation in organic chemistry. *Annals of the New York Academy of Sciences, 988*(1), 141–153. https://doi.org/10.1111/j.1749-6632.2003.tb06093.x

Graulich, N., Hedtrich, S., & Harzenetter, R. (2019). Explicit *versus* implicit similarity—Exploring relational conceptual understanding in organic chemistry. *Chemistry Education Research and Practice, 20*(4), 924–936. https://doi.org/10.1039/C9RP00054B

Haas, D. B., Watts, F. M., Dood, A. J., & Shultz, G. V. (2024). Analysis of organic chemistry students' developing reasoning elicited by a scaffolded case comparison activity. *Chemistry Education Research and Practice*. https://doi.org/10.1039/D4RP00021H

Hayes, J. R. (1996). A new framework for understanding cognition and affect in writing. In C. M. Levy & S. Ransdell (Eds.), *The science of writing: Theories, methods, individual differences, and applications* (pp. 1–27). Lawrence Erbaum Associates.

Houchlei, S. K., Bloch, R. R., & Cooper, M. M. (2021). Mechanisms, models, and explanations: Analyzing the mechanistic paths students take to reach a product for familiar and unfamiliar organic reactions. *Journal of Chemical Education, 98*(9), 2751–2764. https://doi.org/10.1021/acs.jchemed.1c00099

Keiner, L., & Graulich, N. (2020). Transitions between representational levels: Characterization of organic chemistry students' mechanistic features when reasoning about laboratory work-up procedures. *Chemistry Education Research and Practice, 21*(1), 469–482. https://doi.org/10.1039/C9RP00241C

Keiner, L., & Graulich, N. (2021). Beyond the beaker: Students' use of a scaffold to connect observations with the particle level in the organic chemistry laboratory. *Chemistry Education Research and Practice, 22*(1), 146–163. https://doi.org/10.1039/D0RP00206B

Machamer, P., Darden, L., & Craver, C. F. (2000). Thinking about mechanisms. *Philosophy of Science, 67*(1), 1–25.

Martin, P. P., & Graulich, N. (2023). When a machine detects student reasoning: A review of machine learning-based formative assessment of mechanistic reasoning. *Chemistry Education Research and Practice, 24*(2), 407–427. https://doi.org/10.1039/D2RP00287F

Martin, P. P., & Graulich, N. (2024). Beyond language barriers: Allowing multiple languages in postsecondary chemistry classes through multilingual machine learning. *Journal of Science Education and Technology, 33*(3), 333–348. https://doi.org/10.1007/s10956-023-10087-4

Martin, P. P., Kranz, D., Wulff, P., & Graulich, N. (2023). Exploring new depths: Applying machine learning for the analysis of student argumentation in chemistry. *Journal of Research in Science Teaching*. https://doi.org/10.1002/tea.21903

Moon, A., Gere, A. R., & Shultz, G. V. (2018). Writing in the STEM classroom: Faculty conceptions of writing and its role in the undergraduate classroom. *Science Education, 102*(5), 1007–1028. https://doi.org/10.1002/sce.21454

Morin, S., Legault, R., Laliberté, F., Bakk, Z., Giguère, C.-É., de la Sablonnière, R., & Lacourse, É. (2024). StepMix: A python package for pseudo-likelihood estimation of generalized mixture models with external variables. *arXiv*. (arXiv:2304.03853). http://arxiv.org/abs/2304.03853

National Research Council. (2012). *A framework for K-12 science education practices, cross- cutting concepts, and core ideas*. The National Academies Press.

Noyes, K., McKay, R. L., Neumann, M., Haudek, K. C., & Cooper, M. M. (2020). Developing computer resources to automate analysis of students' explanations of London dispersion forces. *Journal of Chemical Education, 97*(11), 3923–3936. https://doi.org/10.1021/acs.jchemed.0c00445

Russ, R. S., Scherr, R. E., Hammer, D., & Mikeska, J. (2008). Recognizing mechanistic reasoning in student scientific inquiry: A framework for discourse analysis developed from philosophy of science. *Science Education, 92*(3), 499–525. https://doi.org/10.1002/sce.20264

Sevian, H., & Talanquer, V. (2014). Rethinking chemistry: A learning progression on chemical thinking. *Chemistry Education Research and Practice, 15*(1), 10–23. https://doi.org/10.1039/C3RP00111C

Watts, F. M., Schmidt-McCormack, J. A., Wilhelm, C. A., Karlin, A., Sattar, A., Thompson, B. C., Gere, A. R., & Shultz, G. V. (2020). What students write about when students write about mechanisms: Analysis of features present in students' written descriptions of an organic reaction mechanism. *Chemistry Education Research and Practice, 21*(4), 1148–1172. https://doi.org/10.1039/C9RP00185A

Watts, F. M., Dood, A. J., & Shultz, G. V. (2022a). Developing machine learning models for automated analysis of organic chemistry students' written descriptions of organic reaction mechanisms. In N. Graulich & G. Shultz (Eds.), *Student reasoning in organic chemistry* (1st ed., pp. 285–303). The Royal Society of Chemistry. https://doi.org/10.1039/9781839167782-00285

Watts, F. M., Park, G. Y., Petterson, M. N., & Shultz, G. V. (2022b). Considering alternative reaction mechanisms: Students' use of multiple representations to reason about mechanisms for a writing-to-learn assignment. *Chemistry Education Research and Practice, 23*(2), 486–507. https://doi.org/10.1039/D1RP00301A

Watts, F. M., Dood, A. J., & Shultz, G. V. (2023a). Automated, content-focused feedback for a writing-to-learn assignment in an undergraduate organic chemistry course. In *LAK23: 13th international learning analytics and knowledge conference* (pp. 531–537). https://doi.org/10.1145/3576050.3576053

Watts, F. M., Dood, A. J., Shultz, G. V., & Rodriguez, J.-M. G. (2023b). Comparing student and generative artificial intelligence Chatbot responses to organic chemistry writing-to-learn assignments. *Journal of Chemical Education, 100*(10), 3806–3817. https://doi.org/10.1021/acs.jchemed.3c00664

Watts, F. M., Finkenstaedt-Quinn, S. A., & Shultz, G. V. (2024). Examining the role of assignment design and peer review on student responses and revisions to an organic chemistry writing-to-learn assignment. *Chemistry Education Research and Practice*. https://doi.org/10.1039/D4RP00024B

Weller, B. E., Bowen, N. K., & Faubert, S. J. (2020). Latent class analysis: A guide to best practice. *Journal of Black Psychology, 46*(4), 287–311. https://doi.org/10.1177/0095798420930932

Williamson, D. M., Xi, X., & Breyer, F. J. (2012). A framework for evaluation and use of automated scoring. *Educational Measurement: Issues and Practice, 31*(1), 2–13. https://doi.org/10.1111/j.1745-3992.2011.00223.x

Yik, B. J., Dood, A. J., Cruz-Ramírez De Arellano, D., Fields, K. B., & Raker, J. R. (2021). Development of a machine learning-based tool to evaluate correct Lewis acid–base model use in written responses to open-ended formative assessment items. *Chemistry Education Research and Practice, 22*(4), 866–885. https://doi.org/10.1039/D1RP00111F

Yik, B. J., Schreurs, D. G., & Raker, J. R. (2023). Implementation of an R Shiny App for instructors: An automated text analysis formative assessment tool for evaluating Lewis Acid–base model use. *Journal of Chemical Education, 100*(8), 3107–3113. https://doi.org/10.1021/acs.jchemed.3c00400

Zaimi, I., Dood, A. J., & Shultz, G. V. (2024). The evolution of an assignment: How a writing-to-learn assignment's design shapes organic chemistry students' elaborations on reaction mechanisms. *Chemistry Education Research and Practice, 25*(1), 327–342. https://doi.org/10.1039/D3RP00197K

Zhai, X., Yin, Y., Pellegrino, J. W., Haudek, K. C., & Shi, L. (2020). Applying machine learning in science assessment: A systematic review. *Studies in Science Education, 56*(1), 111–151. https://doi.org/10.1080/03057267.2020.1735757

Chapter 2
Students' Teleological Reasoning: An Explanatory Escape

Friederike Trommler and Marcus Hammann

2.1 Introduction

In biology education, students struggle to develop adequate mechanistic explanations for natural phenomena. Failure to do so frequently results in teleological reasoning, which attributes function or purpose etiologically to these phenomena. This paper characterizes the phenomenon of teleological reasoning as an *explanatory escape* when students knowingly or unknowingly encounter knowledge gaps and take recourse to functions rather than mechanisms. By understanding this cause of students' teleological reasoning teachers can provide guidance and help students in creating adequate mechanistic explanations.

This chapter is organized into four sections. The first section *Mechanistic Reasoning in Scientific Practice* defines mechanistic reasoning and describes reasoning strategies that professional scientists apply. In the second section, we focus on students' reasoning and explore the extent to which contexts influence students' reasoning. The third section *Students' Reasoning and Explanatory Escape: A qualitative study* explores teleological reasoning as a compensatory mechanism when students lack or cannot access relevant knowledge. Finally, the fourth section *Implication: Guiding Students Towards Mechanistic Reasoning*, draws conclusions

Supplementary Information The online version contains supplementary material available at https://doi.org/10.1007/978-3-032-10246-1_2.

F. Trommler (✉)
Wilhelm-Ostwald-Gymnasium Leipzig, Leipzig, Germany
e-mail: trommler@uni-muenster.de

M. Hammann
Centre for Biology Education, University of Münster, Münster, Germany

N. Graulich et al. (eds.), *Exploring Mechanistic Reasoning in Science Education*, Contributions from Science Education Research 17,
https://doi.org/10.1007/978-3-032-10246-1_2

on how teachers can prevent explanatory-gap situations to overcome students' taking recourse to teleological reasoning and how to develop adequate mechanistic reasoning instead.

2.2 Mechanistic Reasoning in Scientific Practice

In scientific practice, mechanistic reasoning, as described by Machamer et al. (2000), focuses on "entities and activities organized such that they are productive of regular changes from start or set-up to finish or termination conditions"(Machamer et al., 2000). This definition can be unpacked in the following way: Activities produce change. Entities are the things that participate in activities. In the case of the natural selection mechanism,[1] entities are organisms, genes, and the environment, and activities are reproducing, mutating, selecting, and inheriting. Furthermore, this definition can be applied to the example of how the giraffe's long neck evolved: The natural selection mechanism starts with a population of giraffes that vary genotypically and phenotypically in terms of neck length.[2] These giraffes show a range of activities, for example feeding and reproducing. The giraffes eat leaves and live in areas where most leaves are on higher branches. The long-necked giraffes have survival advantages as they can reach more food. Hence long-necked giraffes live longer, are better nourished, stronger, more successful in rival fights, better able to defend their young, females produce more milk, and thus long-necked giraffes reproduce also more frequently. Over several generations, the proportion of long-necked giraffes increases, which leads eventually to the predominance of long-necked giraffes.

Scientists often use various reasoning strategies to construct these mechanistic explanations, such as analogical reasoning, and forward and backward chaining (Darden, 2002). Chaining uses available knowledge to make claims about what must have happened previously to bring about the current state (backward chaining) or what will happen next given certain entities or activities that are present now (forward chaining). By knowing the general properties of entities involved, much can be said about the activities that must have produced them and about the activities in which they can engage. Similarly, examining the characteristics of an activity can provide clues about the entities involved in it and the entities that the activity produced (Darden & Craver, 2002).

[1] Natural selection qualifies as a mechanism according to the definition by Machamer, Darden and Craver if it is instantiated in a specific population with particular environmental conditions. The general template (Darden speaks of a schema) outlines the general principles variation, competition, differential survival, reproduction, and inheritance.

[2] An antecedent mechanism which is not in focus in this context caused the variation in neck length. Where to start and end the mechanism contemplation is essentially the choice of the researcher.

2.3 Students' Reasoning and Its Context-Sensitivity

Students' reasoning in biology often mirrors the strategies professional scientists use to generate mechanistic explanations, albeit with significant differences due to their limited knowledge and experience (Russ, 2006). Russ demonstrates that the language of Machamer, Darden and Craver about mechanistic reasoning can be applied to describe students' reasoning. Mechanistic reasoning is shown from elementary students to introductory college students. Depending on the inquiry context, the students' discourse varies with respect to the elements of mechanistic reasoning that occurs (Russ, 2006). Typically, the description of the target phenomenon is given by the teacher's question. Then, even elementary students, who respond to the question, identify entities and activities and try to causally connect them through chaining. When students chain backward they answer questions along the lines of "What entities were necessary for this activity?" or "What activities could have occurred to result in entities with such properties"? Alternatively, when they chain forward, they ask "What activities can be expected given the fact that there are entities with specific properties?" or "How can this activity affect the surrounding entities?". In contrast to scientists, children's mechanistic reasoning is episodic or fragmented: Sometimes their inquiry can be described along the lines of the mechanism definition by Machamer, Darden and Craver and they reason "about one part of a mechanism on the basis of what is known or conjectured about other parts" (Darden, 2002). At other times, students might fail to identify entities or activities and they cannot generate a mechanistic explanation. Russ concludes that judging students' mechanistic reasoning based on a snapshot could misrepresent their full potential because students' reasoning is highly context-sensitive and depends on the mechanistic knowledge available in the context.

Context and knowledge shape reasoning (Järvilehto, 2015; Gouvea & Simon, 2018). When students encounter science problems, a vast array of information competes for their attention. Perceptual filters determine what students explicitly perceive from implicit inputs. In selecting meaningful over meaningless information, students rely on ideas from previous experience in their everyday life or in the classroom. Besides, they gather available information and make generalizations in search for explanations. Hence, the contextual clues available to students in situations in which they attempt to generate mechanistic explanations affect students' reasoning. Furthermore, students tend to focus on the most salient information (Grotzer & Tutwiler, 2014; Järvilehto, 2015).

Salience is the property by which certain aspects stand out or become more noticeable than others. Salient information captures attention and becomes the basis for intuitive reasoning. Salience results from the interaction between an observer's prior knowledge and experience (top-down salience) and task features (bottom-up or stimulus-driven salience) (Osman & Stavy, 2006). In the context of teleological reasoning the research focus has been on top-down-salience. Kelemen coined the term *teleological mind* to describe the intuitive tendency of people to seek purpose behind things and events (Kelemen, 2012). Besides cognitive predispositions, repeated exposure to certain stimuli contributes to top-down salience. For example, in biology class functions of biological traits are often highlighted so that students' knowledge base and their expectations make them more likely to default to explaining phenomena in terms of function. Furthermore, biological functions such as the appetite-suppressing effect of eating or the light-shielding effect of blinking are deeply tied to our experiences so that functions are prominent in our reasoning. Biological function is also salient in terms of stimulus-driven salience. Particularly noticeable about living beings is their causal organization: the parts of living-beings work together to support each other. Thus, people are prompted to ask what purpose a part serves in an organism (Keil, 1992, 1995). Also, biological functions are often physically perceptible and thus particularly salient because it is a very immediately observable and noticeable outcome of a biological phenomenon. When a plant bends towards a light source, for example, this movement is easily accessible and memorable to observers. In contrast, the underlying mechanism, such as the asymmetric distribution of the plant hormone auxin and the differential growth rates on the light-facing and shaded sides of the plant, are not directly visible and not as salient. Essentially, an oberver's prior knowledge and experience (top-down salience) and task features (stimulus-driven salience) interact and can reinforce each other.

Intuitive reasoning that relies on readily available contextual cues is not per se inadequate; in many situations it is rather pragmatic. If your phone does not charge you think first about common recent causes such as a faulty charging cable or a dirty charging port. You might try a different cable and clean the charging port. This quick reasoning often leads to a solution without delving into more complex mechanical diagnostics initially. Simplifying the complexity of the world around us can be very useful in our daily life because it often results in the right answers with less mental effort (Talanquer, 2006). However, the phone charging issue might be due to less plausible causes so that intuitive reasoning is not successful. Students often bear such a fate in science classes when they look for the most plausible cause based on salient information. In response to the task "Explain how giraffes have developed longer necks than their ancestors" students often answer "Because with the long neck they can reach leaves located higher up." Seizing the readily-available information that the neck is long students jump straightforwardly to the conclusion that this functional feature is a plausible difference-maker that explains the evolutionary change, thereby neglecting the complex evolutionary mechanism.

2.4 Students' Reasoning and Explanatory Escapes: A Qualitative Study

2.4.1 Context and Participants

When students try to generate mechanistic explanations in the absence of knowledge, they tend to take recourse to teleological reasoning. To characterise this cause of students' teleological reasoning we use the term *explanatory escape* first used by van Mil (2013) in connection with teleological reasoning. He argues that people who rely on intentional and anthropomorphic explanations do not necessarily assume actual intentional and anthropomorphic agency, but that they use them as an "explanatory escape" because no "intelligible and plausible underlying mechanism is available".

In biology education, discussions have often placed greater emphasis on the problematic entanglement of teleology with intentionality, while the role of insufficient knowledge as a root cause of teleological reasoning has received comparatively little attention. Research findings from cognitive psychology, in contrast, supports van Mil's characterization of teleological reasoning as an *explanatory escape* through dual-process theories of thinking. According to these theories human thinking operates through two systems: System 1, which is fast, automatic, and relies on intuition, and System 2, which is slow, deliberate, and analytical. When students cannot activate sufficient knowledge in a certain context to engage System 2 for generating mechanistic explanations, they default to System 1, using intuitive reasoning as an explanatory escape.

The specific nature of students' intuitive teleological reasoning in such explanatory escape situations—regardless whether it relies on intentionality or not—remains a topic of debate also among cognitive psychologists. Some researchers argue that teleological reasoning stems from a naïve psychology and thus assumes an intentional agent that strives for purpose. Others suggest that it stems from a naïve biology and thus does not assume intentionality, but need-oriented biological mechanisms (Evans, 2008; Keil, 1995; Kelemen, 2012). Furthermore, cultural beliefs, particularly creationism, are often discussed as contributing factors to teleological reasoning (Evans, 2008; Evans et al., 2010).

To test van Mil's hypothesis that students do not assume intentional agency we conducted personal interviews with 23 students (11 lower (10–13 years), 12 upper (18–19 years) secondary level). We used an evolutionary context since students' teleological explanations are especially prominent in this context. Also, evolutionary tasks often have a strong stimulus-driven salience on function, the effect of which we would like to explore in the interviews. Moreover, lower secondary students commonly lack evolutionary knowledge until their tenth year of schooling so that we expected a high frequency of intuitive teleological reasoning.

2.4.2 Data Collection

This study applied a qualitative research approach to elicit students' reasoning about teleological agency. Students were interviewed individually following standardized interview guidelines. Each interview lasted 25–35 min and consisted of three parts, which dealt with the context of the evolution of the giraffe neck:

1. **Open response task**: To gain initial insights into students' views of the evolution of the giraffe neck, we first asked them to explain in their own words how the long neck evolved. Students completed the first task in written form so that their initial explanation could be used as basis for discussion at the end of the interview

2. **Explanation judgement task**: To explore students' views of teleological agency, students orally indicated their agreement or disagreement with five teleological agent explanations and a natural selection explanation and justified their explanation judgement. The justification is important because mere acceptance of teleological explanations does not necessarily imply teleological reasoning (Zohar & Ginossar, 1998; Trommler et al., 2018; Trommler & Hammann, 2020). The five teleological agent explanations portrayed teleological agents that have been described in the literature based on students' explanatory formulations:

   (a) organisms themselves who adapt to their environment (Baalmann et al., 2004; Bishop & Anderson, 1990; Clough & Wood-Robinson, 1985; Hallden, 1988; Jensen & Finley, 1995; Kampourakis & Zogza, 2008; Moore et al., 2002; Settlage, 1994)
   (b) organisms that decide to reproduce with conspecifics that have favourable traits (Baalmann et al., 2004; Weitzel, 2006)
   (c) Mother Nature (Gregory, 2009; Hallden, 1988; Kampourakis & Zogza, 2008; Kelemen, 2012; Moore et al., 2002)
   (d) the body of organisms (Baalmann et al., 2004; Deadman & Kelly, 1978; Weitzel, 2006)
   (e) God (Kampourakis & Zogza, 2008; Kelemen & Di Yanni, 2005).

The six explanations were presented consistently across participants with the natural selection explanation intentionally positioned as the second-to-last explanation to minimize knowledge-introducing sequencing effects, ensuring that it did not serve as a source of evolutionary knowledge early in the reasoning process. The God explanation appeared last due to its presumed stability based on cultural identity, making it less susceptible to the influence of preceding explanations. To communicate the literal meaning of teleological agency as clearly as possible,[3] the explanation prompts were not only presented orally and in writing by the interviewer, but also illustrated by a cartoon. Consult the supplemental material for the six explanations and the accompanying cartoon drawing.

3. **Final explanation choice task**: To survey students' views after reviewing all explanations, students were asked to name their preferred explanation orally. Moreover, to reveal what aspects of the presented explanations might cause students to change their views on the evolution of the giraffe's neck, students were asked whether they would formulate the same explanation as at the beginning or what other explanation they would give at the end of the interview.

[3] If students'attention is not explicitly drawn to intentional teleology, research shows that they focus on a variety of other features than teleology when judging teleological explanations, such as familiarity, relevance, and complexity of the teleological explanation instead (Trommler et al., 2018). Moreover, without visual aids to clarify the literal intentional meaning of teleology, students might interpret teleological formulations as acceptable shortcuts for the more complex evolutionary mechanism (Trommler & Hammann, 2020). This interpretation would not be surprising, as pedagogical shortcut formulations are part of students'school experience, not only in biology but also in physics and chemistry (Brock & Kampourakis, 2023; Talanquer, 2024; Zohar & Ginossar, 1998).

Congruent with van Mil's (2013) hypothesis we expected patterns across the three interview tasks indicating that students use teleological reasoning as an explanatory escape rather than as an expression of a firm belief in intentional agency. The open response task in itself cannot clearly show belief or disbelief in teleological agency since teleological explanations may reflect as-if-phrasing rather than genuine assumptions of agency (Zohar & Ginossar, 1998). The open response task can provide a first indicator of explanatory escapes in case teleological reasoning correlates with gaps in evolutionary knowledge. The explanation judgement tasks serve as a clearer indicator for students' (dis)agreement with literal teleological agent explanations. At the same time, their reasoning strategies, such as drawing on task features, using everyday knowledge, or expressing uncertainty, may point to teleological reasoning as a tool to bridge explanatory gaps. The final explanation choice task tests the malleability of students' reasoning in the course of the interview. It might refute or support van Mil's hypothesis that teleological reasoning serves as a flexible strategy to address gaps in students' understanding rather than an expression of a firm belief in intentional agency.

2.4.3 Data Analysis

The transcribed interviews were analyzed qualitatively using a combined deductive and inductive approach (Mayring, 2010). Based on the coding scheme of Nehm et al. (2010), students' open responses were coded for the presence or absence of evolutionary key concepts as well as contextually inadequate ideas. Students' explanation judgements, which indicated either agreement or disagreement with the presented explanations, were binary coded. To code students' justifications for their explanation judgements a respective coding schema was developed: Based on a literature review on teleological agency, scoring students' justifications for their explanation judgements started with the deductive assumptions that students might either (a) give intentional teleological justifications that are in accordance with the intentional portrayal of the teleological agents, (b) give need-based teleological justifications that object to the intentional portrayal of the teleological agents, but argue for a need-based operation of the mechanism leading to the neck elongation or (c) give adequate justifications that object to the intentional portrayal of the teleological agents. Inductive data analysis substantiated these deductive assumptions with respect to the individual explanations (for detailed descriptions of the categories consult the supplemental material). Students' final explanation choice was coded as agreement with one of the six presented explanations. The transcripts were coded using the MAXQDA software package, which allowed to record the frequency of the assigned codes across tasks and students. These frequencies are depicted in the results section to demonstrate the prevalence of assigned codes in the data, but they are not meant to imply statistical meaning, as the present research is of a qualitative nature.

2.4.4 Main Findings

Figure 2.1 gives an overview of the categories that describe students' reasoning throughout the interview. The categories are depicted following the course of the interview: On the left Fig. 2.1 depicts the evolutionary key concepts and scientifically inadequate conceptions in students' prior explanations in the open response task, followed by the justifications for students' agreement/disagreement in the middle, and students' final choice once they had been presented all explanations on the far right. In the following paragraphs, we give an overview of students' interview responses (for more detailed descriptions of all categories and exemplary student responses, see the coding guide).

2.4.5 Open Response Task

Students' initial explanations of the evolution of the giraffe's neck revealed a marked contrast between those who possessed evolutionary knowledge and those who did not. Notably, students who incorporated evolutionary key concepts (first column in Fig. 2.1) did not include any teleological ideas (second column in Fig. 2.1), even when their explanations did not include all evolutionary key concepts. Lara, for example, referred merely to differential survival and phenotypic variation and missed differential fitness, heredity, and genotypic variation, but excluded teleological explanatory elements. In contrast, lower secondary students did not incorporate evolutionay key concepts, but teleological ideas instead. For example, Mila argued that "over the course of the centuries the vegetation on earth and the giraffes' feed sources had changed so that the giraffes had to *adapt*." Dana framed nature as an intentional agent by arguing that "As some leaves were too high, they could not reach these leaves with their short neck. Then, *nature was very clever and has made the giraffes' neck grow* over time." Leni implicitly invoked the idea of a purposefully acting body by arguing that "The trees were all very high. To survive, *the spine and neck grew*. Eventually, they reached the juicy treetops." God was not mentioned as a teleological agent by any of the students in the sample.

2.4.6 Explanation Judgement Task

Whereas some lower secondary students agreed with some intentional agency explanations, the majority of students disagreed with them. All upper secondary students agreed with the natural selection explanation, whereas only half of the lower secondary students agreed with it (see column 11 and 12 in Fig. 2.1).

Students' justifications against intentional agency fall into the major categories (a) adequate justifications, and (b) naïve biological justifications. Justifications

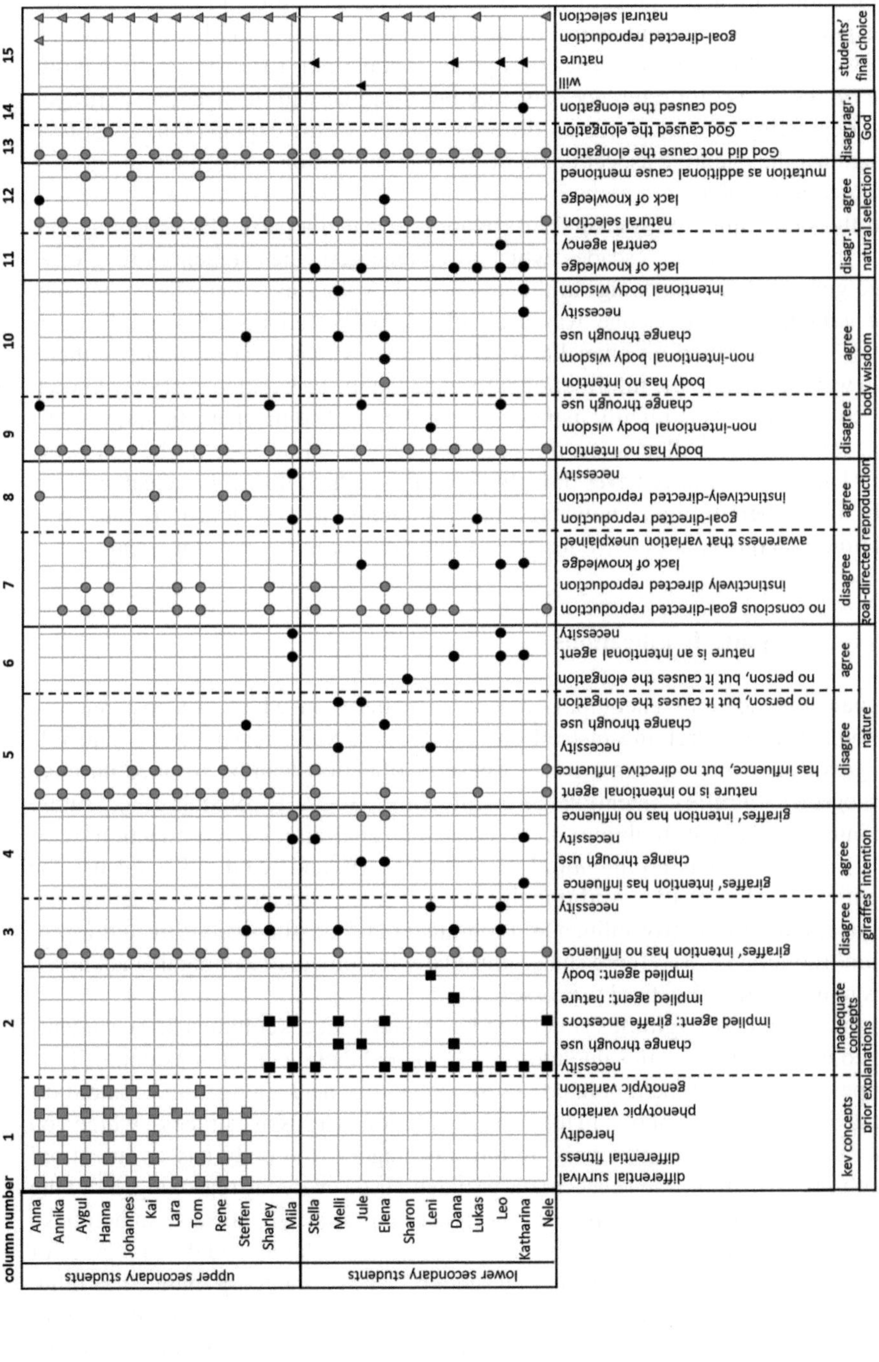

Fig. 2.1 Students' interview responses following the course of the interview: On the far left, the evolutionary key concepts and scientifically inadequate conceptions in students' prior explanation, followed by the justifications for students' agreements in the middle and students' final choice once they had been presented all explanations on the far right. Grey circles indicate adequate justifications, black circles indicate inadequate justifications

supporting teleological agency are subsumed in the category (c) naïve psychological justifications. The students applied many of these justification categories in combination with agree judgements (even-numbered columns in Fig. 2.1) and disagree judgements (odd-numbered columns in Fig. 2.1). This is because students often weighed different properties of a given explanation: for instance, they might agree on the basis of a reference to nature but disagree with its agentive portrayal. Thus, depending on their focus on different aspects of the explanation, a justification against the agentive aspect could occur in combination with either an agree or a disagree judgement.

At the same time, the students' justification revealed a particular reasoning pattern: explicit explanatory escapes, which often emerged when students struggled to bridge gaps in their reasoning. In the absence of evolutionary knowledge, no adequate explanatory knowledge was available, so that students used 'explanatory escapes' as they relied on other knowledge elements that were available to them, which were limited to the information introduced in the task and to everyday knowledge. Some students explicitly mentioned that they lacked mechanistic knowledge. Mark, for example, reflected on missing mechanistic knowledge in the following way: "That's hard. I have never thought about the method how it has changed." On the basis of their limited knowledge base, students drew conclusions that were sometimes adequate, at other times naïve biological, or naïve psychological.

2.4.6.1 Adequate Justifications

Adequate justifications encompass reasoning patterns in which students rejected teleological agency, relying instead on scientifically valid reasoning or plausible analogies from everyday experiences. These justifications can be grouped into five subcategories: 'giraffes' intention has no influence' (Fig. 2.1, columns 3, 4), 'nature is no intentional agent' (column 5), 'no conscious goal-directed reproduction' (column 7), 'the body has no intention' (columns 9, 10), and 'God did not cause the elongation' (column 13). Additionally, two further subcategories—'nature has influence, but no directive influence' (column 5) and 'instinctively-directed reproduction' (columns 7, 8)—represent scientifically legitimate reasoning that acknowledges non-intentional natural processes.

In the subcategory 'giraffes' intention has no influence' (columns 3, 4), students rejected the idea that giraffes intentionally elongated their necks. For example, Tom, an upper secondary student pointed out that a "willful neck elongation" contradicts "what I have learnt at school". In his view, "evolution is a passive process, which you cannot control actively." Similarly, in the subcategory 'the body has no intention' (columns 9, 10), Melli, a lower secondary student, used everyday reasoning. Drawing on her own experience, she explained: "I rather don't think so, because sometimes I also think, 'Well, I would have to be taller to reach something,' and then my neck or my legs also do not start to grow all of a sudden." The subcategory 'nature has influence, but no directive influence' (column 5) comprises responses that are scientifically fully legitimate as they oppose the intentional portrayal of nature, but argue that 'nature' in the sense of 'the ecosystem' did have a—non-directed—influence on the neck elongation. The subcategory 'instinctively-directed reproduction' (columns 7, 8) encompasses reasoning based on the principle of sexual selection as they assume that giraffes do favour conspecifics with longer necks, but that this preference is due to (genetic) predispositions rather than to a conscious decision.

2.4.6.2 Naïve Biological Justifications

Naïve biological justifications encompass reasoning patterns in which students attempt to explain evolutionary change using biologically inspired ideas that are ultimately incorrect or incomplete. These justifications are characterized by their reliance on misconstrued evolutionary concepts, often blending elements of biology with intuitive ad-hoc ideas. Within this main category, several distinct subcategories can be identified: 'necessity' (Fig. 2.1, columns 3–6, 8, 10), 'change through use' (column 3–5, 9, 10), 'nature is no person, but it causes the elongation' (columns 5, 6) and 'non-intentional body wisdom.' (columns 9, 10).

The 'necessity' subcategory (columns 3–6, 8, 10) expresses the idea that the mechanism leading to the neck elongation was guided by the fact that the long neck is necessary for giraffes to reach food and thus to survive. The 'change through use' subcategory (column 3–5, 9, 10) reflects the belief that the giraffe ancestors had stretched towards highly situated leaves and that as a result their necks were elongated over time. Jule exemplified this reasoning by stating, "They stretched to eat up there, so it became stretched. But the giraffe did not really know its neck was becoming stretched." The subcategories 'nature is no person, but it causes the elongation' (columns 5, 6), and 'non-intentional body wisdom' (columns 9, 10) include reasoning where students argue in favour of a *directed* action by nature or the body, but oppose an *intentional* action. For example, Leni invoked nature and development as placeholders for the unknown mechanism: "They cannot say, 'My neck should become longer now!' but through nature and this development, it worked eventually."

2.4.6.3 Naïve Psychological Justifications

Naïve psychological justifications include reasoning patterns in which students supported teleological agency by attributing intentionality, purpose, or conscious decision-making to giraffes, nature, or a higher power. These justifications can be grouped into six subcategories: 'giraffes' intention has influence' (Fig. 2.1, column 4), 'nature is an intentional agent' (column 6), 'goal-directed reproduction' (column 8), 'intentional body wisdom' (column 10), 'central agency' (column 11) and 'God caused the elongation' (columns 13, 14). Each subcategory reflects distinct ways of reasoning, often relying on intuitive purpose-driven explanations.

For example, the subcategory 'giraffes' intention has influence' (column 4) comprises an agreement with a willful neck elongation, which the student justified with the words "it must have happened somehow." In the subcategory 'nature is an intentional agent' (column 6), students interpreted the order and structure observed in nature as evidence of intentional action. For example, Dana argued: "Nature does everything well with all the circles and so on." The subcategory 'goal-directed reproduction' (column 8) includes justifications that assume that giraffes had a conscious goal in their reproductive behaviour, with some justifying their reasoning by pointing to other observable goal-directed actions, such as searching for food. The subcategory 'intentional body wisdom' (column 10) includes reasoning where bodily processes are seen as intentionally directed toward a final goal. Melli exemplified this with an analogy to human growth: "As a baby you grow as well, and you know that you have to be as tall as a normal human, and then all the bones grow, and that's probably how it is with the giraffe." The subcategory 'central agency' (column 11) reflects reasoning where an overarching, conscious force is assumed to have directed the evolutionary change. Leo, for instance, rejected natural selection by stating: "Someone must have done it." Finally, in the subcategory 'God caused the elongation' (columns 13, 14) Katharina based her reasoning on divine intervention, arguing: "He created the world, otherwise they would have died out."

2.4.7 Discussion of Study Results

The results of this study support van Mil's (2013) hypothesis that students often employ teleological reasoning as an explanatory escape rather than as an expression of a firm belief in intentional agency. While most students appear to rely on such explanatory escapes, the presence of agreement with intentional teleological explanations could solidify into firm beliefs in some students if such reasoning is left unchallenged. However, across the three interview tasks, patterns suggest that teleological reasoning frequently serves as an intuitive strategy to bridge explanatory gaps when evolutionary knowledge is incomplete or absent.

In the *Open Response Task*, a clear distinction was observed between students with and without evolutionary knowledge. Students who demonstrated key evolutionary concepts in their explanations rarely included teleological reasoning, even if their explanations were incomplete. For example, Lara's explanation, while omitting elements such as differential fitness and heredity, still avoided teleological assumptions. In contrast, students with limited evolutionary understanding relied on teleological reasoning to construct their explanations. Importantly, God was not mentioned as a causal agent by any student, underscoring that teleological reasoning in the sample did not stem from firmly held religious beliefs.

The *Explanation Judgement Task* provided further insights into how students evaluated intentional agency explanations. Crucially, students' justifications, categorized into *adequate justifications*, *naïve biological justifications*, and *naïve psychological justifications*, featured explicit explanatory escapes, where students attempted to reconcile their lack of knowledge with intuitive reasoning patterns.

Adequate justifications demonstrated the students' ability to reject teleological explanations based on scientifically sound reasoning or plausible analogies from their everyday experiences. On the other hand, *naïve biological justifications* revealed attempts to make sense of evolutionary processes through misconstrued biological concepts. For instance, Jule's assertion that giraffes' necks stretched over time through repeated effort reflects a wrong idea, but not an expression of genuine belief in intentional agency. *Naïve psychological justifications* illustrate how students invoked intentionality, purpose, or agency when grappling with incomplete knowledge. These results are in line with Kelemen's (2012) hypothesis that teleological reasoning is due to naïve psychology. However, the justifications in this category with their searching character in the face of knowledge gaps rather than fully convinced arguments do not reflect deep-seated beliefs in teleological agency.

The *Final Explanation Choice Task* demonstrated the malleability of students' reasoning: Upper secondary students and most lower secondary students shifted towards the natural selection explanation. Even students who retained elements of teleological reasoning integrated new information into their explanatory frameworks. This adaptability further supports the view that teleological reasoning in our sample served as a flexible, context-dependent strategy rather than a fixed belief system.

In conclusion, the results align with van Mil's (2013) hypothesis that teleological reasoning functions as an explanatory escape rather than as evidence of a firm belief in intentional agency. Our results support also the argument by Gouvea & Simon (2018) that teleological reasoning is sensitive to local context rather than due to firmly held beliefs about teleological causality as has been argued by Coley and Tanner (2012) for example. Gouvea and Simon's (2018) broader claim—that reasoning is not the "straightforward expression of structures in the mind; it is an act of construction, building from knowledge that is activated and assembled in particular situations"– is also consistent with our data.

In our study, students appear to rely on teleological reasoning primarily in the face of knowledge gaps, using intuitive explanations as cognitive tools to make sense of complex biological phenomena. These findings emphasize the importance of addressing these reasoning patterns in science education by providing students with robust, scientifically accurate frameworks for understanding evolutionary processes.

2.4.8 Limitations

While this study provides valuable insights into students' teleological reasoning as an explanatory escape, several limitations must be acknowledged. First, the study relies on a relatively small sample size (23 students), which may limit the generalizability of the findings to broader populations. Second, interviews, while allowing to capture students' reasoning, are inherently limited by their reliance on self-reported reasoning, which may not fully capture students' spontaneous thought processes. Third, the sequence of the explanations in the explanation judgement task may have influenced students' reasoning. Additionally, the groups that differed in their evolutionary knowledge—lower versus upper secondary students—differed also in their age so that this could be another contributing factor causing a higher degree of teleological reasoning. Lastly, cultural and educational background factors, which can significantly influence teleological reasoning, were not deeply explored. Future research with larger, more diverse samples and varied biological contexts is needed to address these limitations and validate the findings more broadly.

2.4.9 Conclusion: The Explanatory Escape: A New Perspective on Missing Mechanistic Knowledge

Traditionally, educational research has defined mechanistic explanations by contrasting them with teleological explanations (Abrams & Southerland, 2001; Carey, 1995; Richardson, 1990). This comparison contrasts the "how" with the "why" and the "adequate" scientific explanation with the "inadequate" student answer. For a long time the elements that constitute the mechanistic "how" remained undefined. Then, Machamer et al. (2000) defined mechanistic reasoning as the identification and description of the entities and activities that produce regular changes from start to finish conditions. Furthermore, Russ (2006) showed that the definition can be applied to describe student discourse. Using the vocabulary from Machamer et al. (2000), we argue that students who have been asked to explain the evolutionary change of traits typically fail to identify and describe the correct mechanistic entities and activities, namely the variations between organisms within a population, selection and its impact on differential survival and reproduction rates. Instead, they provide the start and finish conditions of the mechanism and escape to intuitively available information to fill the gap in-between. Students brainstorm based on their experiences to develop meaningful hypotheses, chaining from the setup conditions and entities provided in the task to the finish conditions to find productive continuity (see Fig. 2.2 for an illustrating schema). While their reasoning may be inadequate from a subject-matter-knowledge perspective, their reasoning strategies hold potential for successful scientific inquiry. Students do not necessarily produce teleological explanations that are completely non-mechanistic and wrong in every respect. With proper scaffolding, students can avoid the need to "escape" and develop more accurate mechanistic explanations.

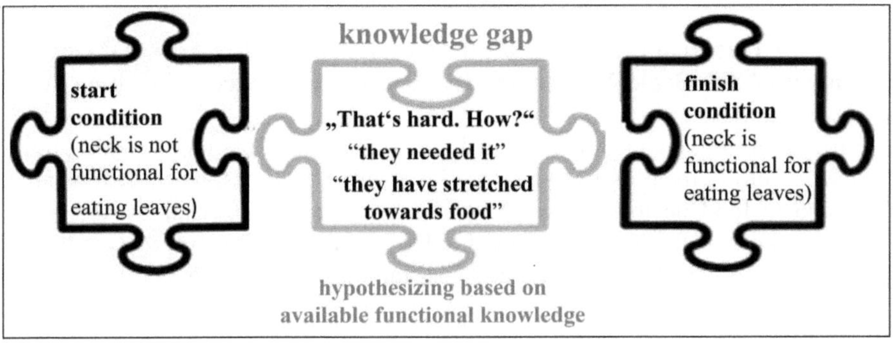

Fig. 2.2 Evolutionary phenomena are often presented in a way that the biological function of phenomena is salient, whereas the evolutionary mechanism is not. Students are thus likely to resort to functional information to close their knowledge gaps

2.5 Implications: Guiding Students Towards Mechanistic Reasoning

Research suggests that teleological reasoning is not merely a result of knowledge gaps but may also be shaped by instructional choices and educational materials (Gouvea & Simon, 2018; Talanquer, 2007). We recommend developing tasks that avoid the misleading potential of stimulus-driven salience of biological function. To help students develop adequate mechanistic explanations, educators should be aware of the insight from the philosophy of biology that biological phenomena can be viewed from two perspectives: mechanistic and functional (for examples see Boxes 2.1 and 2.2). The mechanistic perspective involves reasoning from setup conditions to the finish conditions, while the functional perspective involves reasoning from finish conditions to setup conditions. The perspective influences the direction of students' reasoning and the information provided in the task give cues for their reasoning. To guide students who lack evolutionary knowledge towards mechanistic rather than teleological inferences, we suggest developing the evolutionary mechanism from a mechanistic perspective. This involves starting with a population displaying variation and reconstructing one mechanistic step after the other in a forward-looking direction. Such a forward-looking step-by-step approach provides a logical chain of thought, avoiding explanatory gaps that lead students to resort to teleological reasoning.

The item in Box 2.1 makes the function of elephants' long trunks salient by explicitly stating its utility. This stimulus-driven salience activates the already existing top-down salience, i.e. the cognitive predisposition to reason in terms of function. Confronted with this item students who lack evolutionary knowledge are likely to explain the emergence of the trunk in terms of purpose. In contrast, by introducing variation, the mechanistic reasoning prompt in Box 2.2 focuses students on a

Box 2.1 This Item Is an Example for a Functional Perspective that Cues Students' Teleological Reasoning Due to the Prominence of Function in the Task

"Elephants as we know them, are animals with a remarkably long trunk. This feature allows them to browse on the leaves from the trees, when there is no adequate food on the ground. Nowadays we know that elephants did not always possess this feature but used to have no trunks. Can you explain how the trunks of the elephants emerged?" (Kampourakis & Zogza, 2009).

Box 2.2 This Reasoning Prompt Takes a Mechanistic Perspective

"These animals [referring to a picture] are called piloses. This is how the group of fully-grown adult piloses looked a long time ago, many hundreds of years in the past. They were piloses, but they looked a little bit different from each other. Many of the piloses had wider trunks, and only a small number had thinner trunks." "Here [referring to a picture] are some different fully-grown adult piloses. These are the piloses that live nowadays, and this is what the group looks like now. Piloses nowadays mostly have thinner trunks. Why were there so many piloses with wider trunks in the group a long time ago but nowadays there are mostly piloses with thinner trunks in the group?" (Kelemen & Child Cognition Team, 2017).

mechanistically relevant difference between entities. Starting from organismic variation the inference to the next mechanistic step that organisms differ in their survival capabilities is more likely activated. The prompt involving the fictional piloses comes from the children's book *How the piloses evolved skinny noses*. Using this book Kelemen et al. (2014) explored elementary students' capacities to self-generate the logic of natural selection and concluded that the basic population-based logic of natural selection is within the reach of elementary school-age children.

Further empirical support for the principle of choosing the reasoning direction deliberately comes from Hong et al. (2005) who report that children perform better in cause-effect reasoning tasks than in effect-cause reasoning tasks. By starting the reasoning process with the entities that start further activities and affect other entities children can use the task features as a basis for chaining without any explanatory gaps. Also, van Mil (2013) reports that students struggle to provide adequate explanations when the task requires them to explain cell functions, because they are left to base their chaining on prior knowledge elements. He also describes that tasks that take a mechanistic perspective can successfully prompt students to reason about entities and activities within a cell in order to explain cell functions. Furthermore, in chemistry education Grove et al. (2012) found that mechanistic reasoning in a forward direction advances students in solving organic chemistry problems.

We do not argue for a rigid scaffold as the one and only method for evolution lessons. Rather, we want to raise awareness for the impact of the reasoning direction teachers initiate through tasks and in conversations. We can think our students capable of mechanistic reasoning, most of them (at least in our sample in Germany) do not hold deeply ingrained beliefs about teleological agents, but escape to teleological reasoning when they cannot access knowledge about mechanistic entities and activities. Teachers need to provide students with the building blocks for adequate mechanistic reasoning about evolution. We suggest long-term groundwork by highlighting variation within populations in biological contexts such as vertebrates, or spermatophytes in primary school already. Before students are challenged to reason across multiple mechanistic steps, single chaining steps, such as the stochastic effect of differential survival on the next generation, should be discussed, consolidated and reinforced. As a result, students have reliable stepping stones for the complex reasoning about natural selection.

Besides teacher guidance in structuring students' search for mechanistic explanations, students can benefit from explicit discussions of explanatory gaps. Livni-Alcasid and Haskel-Ittah refer to these gaps as *black boxes* in their chapter *Explicating Black Boxes in the Biology Classroom for Promoting Mechanistic Reasoning* in this book, emphasizing the importance of making students aware of such knowledge gaps in mechanisms. Explicitly addressing black boxes might enhance students' awareness of these gaps and help prevent them from intuitively filling them with teleological reasoning.

The focus of this chapter on the explanatory escape in cases of explanatory gaps represents one aspect that is momentous for evolution education. Ultimately, educators should provide students with a combination of knowledge about the general nature of causality, domain-specific knowledge, and meta-level discussions to direct their intuitive reasoning towards adequate biological reasoning. Knowledge about the general nature of causality includes understanding distributed causality and feedback loops. These concepts can be introduced using everyday phenomena and can be explored in simulations in biological contexts. Meta-level discussions about different explanation types, for example a natural selection explanation in contrast to teleological agent explanations, can clarify the meaning of ambiguous language such as need or adapt. By doing so, students can develop a more comprehensive understanding of biological phenomena and move beyond teleological explanations.

The discussion on teleological reasoning often has a focus on content adequacy, which is one side of the coin in science class. On the other side, students—just like scientists—reason about phenomena and thereby build hypotheses that might not be correct at first. Practices like booing (as described in Bravo & Cofré, 2016) if a student uses the word *need* in describing evolutionary phenomena hampers students' confidence in reasoning. Reasoning based on the knowledge available is a scientific skill that teachers should value so that students are not afraid of wrong answers and instead encouraged to reason independent of social pressure.

Acknowledgments We thank Pia Staudacker for her valuable support in conducting the interviews and Clemens Trommler for creating the cartoons.

References

Abrams, E., & Southerland, S. (2001). The how's and why's of biological change: How learners neglect physical mechanisms in their search for meaning. *International Journal of Science Education, 23*(12), 1271–1281.

Baalmann, W., Frerichs, V., Weitzel, H., Gropengießer, H., & Kattmann, U. (2004). Schülervorstellungen zu Prozessen der Anpassung: Ergebnisse einer Interviewstudie im Rahmen der didaktischen Rekonstruktion. *Zeitschrift für Didaktik der Naturwissenschaften, 10*, 7–28.

Bishop, B. A., & Anderson, C. W. (1990). Student conceptions of natural selection and its role in evolution. *Journal of Research in Science Teaching, 27*(5), 415–427.

Bravo, P., & Cofré, H. (2016). Developing biology teachers' pedagogical content knowledge through learning study: The case of teaching human evolution. *International Journal of Science Education, 38*(16), 2500–2527.

Brock, R., & Kampourakis, K. (2023). A justification of legitimate teleological explanations in physics education. *Science & Education, 32*, 927–945.

Carey, S. (1995). On the origin of causal understanding. In D. Sperber, D. Premack, & A. J. Premack (Eds.), *Causal cognition: A multidisciplinary debate* (pp. 268–308). Clarendon Press.

Clough, E. E., & Wood-Robinson, C. (1985). How secondary students interpret instances of biological adaptation. *Journal of Biological Education, 19*(2), 125–130.

Coley, J. D., & Tanner, K. D. (2012). Common origins of diverse misconceptions: Cognitive principles and the development of biology thinking. *CBE—Life Sciences Education, 11*(3), 209–215.

Darden, L. (2002). Strategies for discovering mechanisms: Schema instantiation, modular subassembly, forward/backward chaining. *Philosophy of Science, 69*(3), 354–365.

Darden, L., & Craver, C. (2002). Strategies in the interfield discovery of the mechanism of protein synthesis. *Studies in History and Philosophy of Science Part C: Studies in History and Philosophy of Biological and Biomedical Sciences, 33*(1), 1–28.

Deadman, J. A., & Kelly, P. J. (1978). What do secondary school boys understand about evolution and heredity before they are taught the topics? *Journal of Biological Education, 12*(1), 7–15.

Evans, E. M. (2008). Conceptual change and evolutionary biology: A developmental analysis. In S. Vosniadou (Ed.), *Educational psychology handbook series. International handbook of research on conceptual change* (pp. 263–294). Routledge.

Evans, E. M., Spiegel, A. N., Gram, W., Frazier, B. N., Tare, M., Thompson, S., & Diamond, J. (2010). A conceptual guide to natural history museum visitors' understanding of evolution. *Journal of Research in Science Teaching, 47*(3), 326–353.

Gouvea, J. S., & Simon, M. R. (2018). Challenging cognitive construals: A dynamic alternative to stable misconceptions. *CBE-Life Sciences Education, 17*(2).

Gregory, T. R. (2009). Understanding natural selection: Essential concepts and common misconceptions. *Evolution: Education and Outreach, 2*(2), 156–175.

Grotzer, T. A., & Tutwiler, M. S. (2014). Simplifying causal complexity: How interactions between modes of causal induction and information availability lead to heuristic driven reasoning. *Mind, Brain, and Education, 8*(3), 97–114.

Grove, N. P., Cooper, M. M., & Cox, E. L. (2012). Does mechanistic thinking improve student success in organic chemistry? *Journal of Chemical Education, 89*(7), 850–853.

Hallden, O. (1988). The evolution of the species: Pupil perspectives and school perspectives. *International Journal of Science Education, 10*(5), 541–552.

Hong, L., Chijun, Z., Xuemei, G., Shan, G., & Chongde, L. (2005). The influence of complexity and reasoning direction on children's causal reasoning. *Cognitive Development, 20*(1), 87–101.

Järvilehto, L. (2015). *The nature and function of intuitive thought and decision making*. Springer.

Jensen, M. S., & Finley, F. N. (1995). Teaching evolution using historical arguments in a conceptual change strategy. *Science Education, 79*(2), 147–166.

Kampourakis, K., & Zogza, V. (2008). Students' intuitive explanations of the causes of homologies and adaptations. *Science & Education, 17*(1), 27–47.

Kampourakis, K., & Zogza, V. (2009). Preliminary evolutionary explanations: A basic framework for conceptual change and explanatory coherence in evolution. *Science & Education, 18*(10), 1313–1340.

Keil, F. C. (1992). The emergence of an autonomous biology. In M. Gunnar & M. Maratsos (Eds.), *Modularity and constraints in language and cognition: The Minnesota symposia* (pp. 103–138). Erlbaum.

Keil, F. C. (1995). The growth of causal understandings of natural kinds. In D. Sperber, D. Premack, & A. J. Premack (Eds.), *Symposia of the Fyssen Foundation. Causal cognition: A multidisciplinary debate* (pp. 234–267). Clarendon Press/Oxford University Press.

Kelemen, D. (2012). Teleological minds: How natural intuitions about agency and purpose influence learning about evolution. In K. Rosengren, S. K. Brem, E. M. Evans, & G. M. Sinatra (Eds.), *Evolution challenges: Integrating research and practice in teaching and learning about evolution* (pp. 66–92). Oxford University Press.

Kelemen, D., & Child Cognition Team. (2017). *How the piloses evolved skinny noses*. Tumblehome Learning.

Kelemen, D., & Di Yanni, C. (2005). Intuitions about origins: Purpose and design in children's reasoning about nature. *Journal of Cognition and Development, 6*(1), 3–31.

Kelemen, D., Emmons, N. A., Seston Schillaci, R., & Ganea, P. A. (2014). Young children can be taught basic natural selection using a picture-storybook intervention. *Psychological Science, 25*(4), 893–902.

Machamer, P., Darden, L., & Craver, C. F. (2000). Thinking about mechanisms. *Philosophy of Science, 67*(1), 1–25.

Mayring, P. (2010). *Qualitative inhaltsanalyse: Grundlagen und techniken* [Qualitative content analysis: Basics and techniques] (11th ed.). Studium Paedagogik. Weinheim: Beltz.

Moore, R., Mitchell, G., Bally, R., Inglis, M., Day, J., & Jacobs, D. (2002). Undergraduates' understanding of evolution: Ascriptions of agency as a problem for student learning. *Journal of Biological Education, 36*(2), 65–71.

Nehm, R. H., Ha, M., Rector, M., Opfer, J., Perrin, l., Ridgway, J., & Mollohan, K. (2010). *Scoring guide for the open response instrument (ORI) and evolutionary gain and loss test (ACORNS).* Technical report of National Science Foundation REESE Project 0909999.

Osman, M., & Stavy, R. (2006). Development of intuitive rules: Evaluating the application of the dual-system framework to understanding children's intuitive reasoning. *Psychonomic Bulletin & Review, 13*(6), 935–953.

Richardson, D. R. (1990). A survey of students' notions of body function as teleologic or mechanistic. *American Journal of Physiology, 258*(6), 8–10.

Russ, R. S. (2006). *A framework for recognizing mechanistic reasoning in student scientific inquiry.* Doctoral dissertation. University of Maryland.

Settlage, J. (1994). Conceptions of natural selection: A snapshot of the sense-making process. *Journal of Research in Science Teaching, 31*(5), 449–457.

Talanquer, V. (2006). Commonsense chemistry: A model for understanding students' alternative conceptions. *Journal of Chemical Education, 84*(5), 811–816.

Talanquer, V. (2007). Explanations and teleology in chemistry education. *International Journal of Science Education, 29*(7), 853–870.

Talanquer, V. (2024). How we simplify, circumvent, or distort causal mechanistic reasoning in chemistry. *Journal of Chemical Education, 101*, 1793–1797.

Trommler, F., & Hammann, M. (2020). The relationship between biological function and teleology: Implications for biology education. *Evolution: Education and Outreach, 13*(11), 1–16.

Trommler, F., Gresch, H., & Hammann, M. (2018). Students' reasons for preferring teleological explanations. *International Journal of Science Education, 40*(2), 159–187.

van Mil, M. (2013). *Learning and teaching the molecular basis of life.* Doctoral dissertation. Universiteit Utrecht.

Weitzel, H. (2006). *Biologie verstehen: Vorstellungen zu Anpassung.* Doctoral dissertation. Universität Hannover.

Zohar, A., & Ginossar, S. (1998). Lifting the taboo regarding teleology and anthropomorphism in biology education - heretical suggestions. *Science Education, 82*, 679–697

Chapter 3
Mechanistic Reasoning About Alternatives: How Students Reason About Plausible and Implausible Mechanisms in Chemistry

Paul P. Martin, Brandon J. Yik, Benjamin T. Burlingham, and Nicole Graulich

3.1 Introduction

Mechanisms of phenomena can be conceptualized as a dynamic interplay of entities and activities organized in a way to rationalize consistent changes from start to finish conditions (Machamer et al., 2000). Given this conceptualization, every mechanism contains four basic features: a phenomenon requiring explanation, the involved entities, the properties of those entities, and the resulting activities (Craver, 2007; Illari & Williamson, 2012). In organic chemistry, explaining phenomena encompasses how and why chemical reactions occur. Here, entities are the atoms and molecules involved in a mechanism, which are defined by specific properties, including their atom types, connectivity, spatial orientation, and electron distribution. Activities refer to the dynamic processes that entities exhibit during a mechanism, for instance, a nucleophilic substitution. A primary driving force for initiating these activities is to attain a lower energy state through bond rearrangements. Consequently, mechanisms in organic chemistry provide a model of continuous electron flow to illustrate, describe, explain, or control the underlying processes that drive chemical reactions (Goodwin, 2003, 2012).

Reasoning about mechanisms is crucial as it allows chemists to anticipate the selectivity of transformations, consider alternative pathways, design novel synthesis

P. P. Martin · N. Graulich (✉)
Institute of Chemistry Education, Justus-Liebig-University, Giessen, Germany
e-mail: Nicole.Graulich@didaktik.chemie.uni-giessen.de

B. J. Yik
Department of Chemistry, University of Georgia, Athens, GA, USA

B. T. Burlingham
Department of Chemistry, Indiana University, Bloomington, IN, USA

N. Graulich et al. (eds.), *Exploring Mechanistic Reasoning in Science Education*, Contributions from Science Education Research 17,
https://doi.org/10.1007/978-3-032-10246-1_3

routes, and critically evaluate proposed reactions (Bhattacharyya & Bodner, 2005). Given these affordances, instruction must explicitly support chemistry students' mechanistic reasoning so they can consider multiple molecular properties to predict the plausibility of various products (Grove & Bretz, 2010). This explicit support might help students move beyond a simplistic view of mechanisms as a linear sequence of events; instead, students can learn to integrate the complex interactions of entities and activities in their mechanistic reasoning (Weinrich & Sevian, 2017).

When dealing with mechanisms of varying plausibility, students need additional support to handle the arising ambiguity in the reasoning process (Cruz-Ramírez de Arellano & Towns, 2014; Grove & Bretz, 2010; Lieber et al., 2022a, b). Therefore, instruction should guide students to shift their perspective from believing that a mechanism yields only one defined product to understanding that reagents can produce various products and byproducts with differing probabilities (Grove & Bretz, 2010). This shift could be essential as it can contribute to a more sophisticated perception of a mechanism as an outcome of interrelated interactions constrained by energy and entropy, rather than a pre-defined outcome (Talanquer, 2013). Hence, this study investigates differences in students' mechanistic reasoning complexity when judging the plausibility of alternative mechanisms. By evaluating students' reasoning about alternative mechanisms, we aim to support their mechanistic reasoning as well as their understanding that multiple outcomes of a mechanism can coexist in future formative assessments.

3.2 Conceptual Background

Mechanistic reasoning can be conceptualized as a form of causal reasoning that explains the sequential stages, from input to output, of the underlying causal events leading to a phenomenon (Krist et al., 2019; Machamer et al., 2000; Russ et al., 2008). Consequently, mechanistic reasoning extends beyond simply describing a phenomenon. It offers a theoretical perspective on the regularities underlying empirical observations, aiming to propose and reason through possible mechanisms that provide testable hypotheses on how and why phenomena occur (Darden & Craver, 2002; Russ et al., 2008). By systematically identifying, unpacking, and linking the underlying properties that impact entities and activities, mechanistic reasoning is a valuable strategy for explaining or controlling the outcome of scientific phenomena (Krist et al., 2019).

This study builds upon the *essential epistemic heuristics for mechanistic reasoning* by Krist et al. (2019) to categorize undergraduate organic chemistry students' reasoning about alternative mechanisms (Fig. 3.1). Krist et al. (2019) specify three essential epistemic heuristics for evaluating the extent of students' engagement in mechanistic reasoning. The first essential epistemic heuristic involves reasoning across different scalar levels, especially considering the scalar level below the target phenomenon (Krist et al., 2019). In other words, reasoning about what is happening at a scale below the target phenomenon forms the foundation for complex

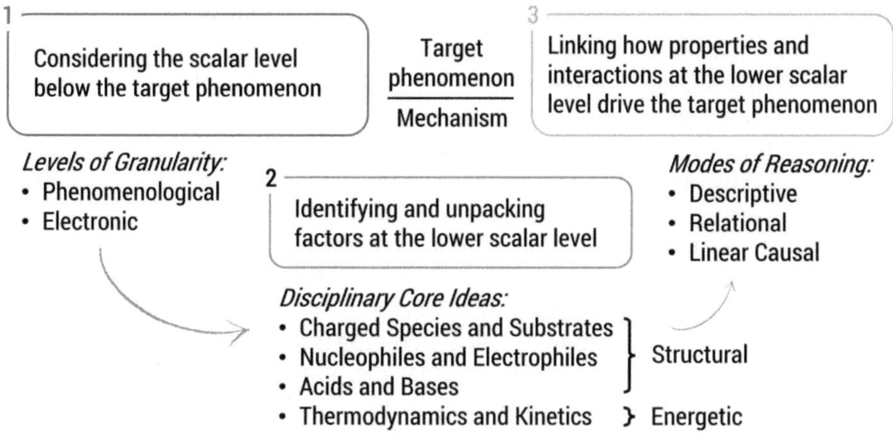

Fig. 3.1 Essential epistemic heuristics (Krist et al., 2019) to classify students' mechanistic reasoning adapted to an organic chemistry context

mechanistic reasoning that characterizes the causes of underlying processes. In our organic chemistry context, we used two *levels of granularity* to classify whether a student considered the scalar level below the target phenomenon (Fig. 3.1): *phenomenological* and *electronic* (Bodé et al., 2019; Deng & Flynn, 2021). *Phenomenological* reasoning includes measurable properties of entities or activities like pK_a values, the direction of chemical equilibrium, or reaction rate, without examining electron distribution. In this way, *phenomenological* reasoning about organic chemistry reactions tends to consider only the target phenomenon while disregarding the lower scalar level. A lower scalar level encompasses *electronic* properties. *Electronic* reasoning discusses the distribution and movement of electrons in chemical compounds through electronic effects, such as resonance, hyperconjugation, or inductive effects. Therefore, the *electronic* level represents a lower scalar level than the *phenomenological* level, making *electronic* reasoning more complex than *phenomenological* reasoning.

Once students examine the lower scalar level, they must identify and unpack the relevant entities, activities, and properties that drive the phenomenon (Krist et al., 2019). Identifying and unpacking relevant entities, activities, and properties requires knowledge of disciplinary core ideas (Grove et al., 2012). Our analysis used disciplinary core ideas related to *structural* and *energetic* features to determine the entities and activities students reasoned about (Bodé et al., 2019; Deng & Flynn, 2021). Disciplinary core ideas related to *structural* features mainly concentrate on the properties of entities. In chemistry contexts, *structural* features like atom connectivity, atom sizes, or charges govern the properties of entities. To categorize the entities and their corresponding properties in students' mechanistic reasoning, we disaggregated this *structural* level into three disciplinary core ideas according to the properties students assigned to the entities: *charged species and substrates*, *nucleophiles and electrophiles*, and *acids and bases* (Martin et al., 2024). Instead, disciplinary core ideas related to *energetic* features mainly focus on the properties of activities. In chemistry, these properties of activities include thermodynamic and kinetic changes like the reaction rate, any influence of entropic effects on the reaction process, and whether a reaction favors reactants or products. We introduced the disciplinary core idea of *thermodynamics and kinetics* to capture students' *energetic* mechanistic reasoning (Fig. 3.1). Together, the disciplinary core ideas identify and unpack the properties of entities and activities in students' mechanistic reasoning about chemical reactions.

After identifying and unpacking the relevant entities, activities, and properties, these factors must be organized to form a sequence of causes over time or space. Accordingly, the third essential epistemic heuristic focuses on how well the parts of the underlying mechanism explain the observed phenomenon (Krist et al., 2019). This reasoning requires linking the identified properties to the phenomenon at its original scalar level and examining whether the unpacked entities and activities fully explain the phenomenon. To evaluate the degree of causality in this third heuristic, we applied three *modes of reasoning* (Fig. 3.1): *descriptive*, *relational*, and *linear causal* (Sevian & Talanquer, 2014). *Descriptive* reasoning only identifies explicit properties of single entities without providing any connections between those entities. The next level, *relational* reasoning, starts to unpack the implicit properties of entities, establishes relationships between a constrained number of those entities, and builds correlative links to the target phenomenon. Ultimately, *linear causal* reasoning unpacks the explicit and implicit properties of most entities and creates causal links to the target phenomenon. In sum, these *modes of reasoning* evaluate the causality of students' mechanistic reasoning. We regard *linear causal* reasoning as the most complex form.

Figure 3.2 illustrates how the three essential epistemic heuristics may appear in students' mechanistic reasoning when they judge the plausibility of a mechanism

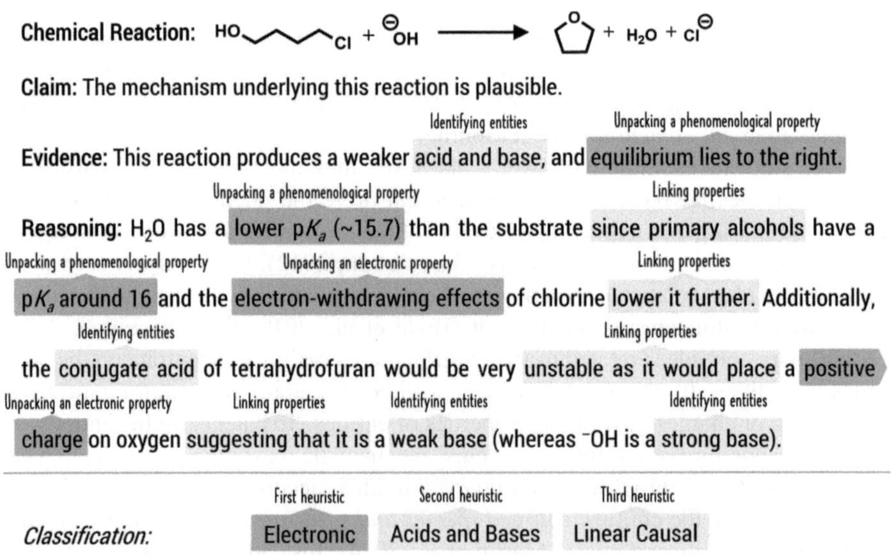

Fig. 3.2 Applying the essential epistemic heuristics to an exemplary student response

within our organic chemistry context (Sect. 3.4). As shown in the student example, reasoning about products and byproducts with varying plausibility may require students to unpack a range of *phenomenological* and *electronic* properties. Combining the *levels of granularity* to operationalize the scalar level of students' mechanistic reasoning in the first heuristic with the *modes of reasoning* to operationalize students' use of causality in the third heuristic served as the foundation for evaluating the complexity of students' mechanistic reasoning. Accordingly, this analysis focuses on students' application of the first and third essential epistemic heuristics. Variations in how students apply the second heuristic are not examined in this article since the use of disciplinary core ideas is context-dependent (Deng & Flynn, 2021). For instance, an acid-base reaction may naturally prompt students to reason about acids and bases, so we hypothesize that students' use of disciplinary core ideas varies substantially depending on the specific mechanism. Consequently, longitudinal comparisons of disciplinary core ideas may only be partially valid and are, thus, not the focus of this article.

3.3 Research Question

Given this conceptualization, it becomes evident that mechanistic reasoning in organic chemistry is a complex practice. Besides, the possibility that the same reagents can produce multiple products of varying plausibility is not straightforward for students during decision-making (Grove & Bretz, 2010). Following this, the complexity of students' mechanistic reasoning may vary depending on their evaluation of a mechanism's plausibility. To identify the characteristics that facilitate complex mechanistic reasoning when confronted with alternatives, we compared the mechanistic reasoning complexity of undergraduate organic chemistry students for multiple sets of alternative mechanisms. Our analysis is guided by the following research question: To what degree does the complexity of students' mechanistic reasoning differ among response characteristics when reasoning about the plausibility of mechanisms? We examined the impact of three response characteristics on students' mechanistic reasoning complexity: the given reaction (Context), the actual plausibility or implausibility of a mechanism (Plausibility), and a student's claim about a mechanism's plausibility (Claim).

3.4 Context of the Study

3.4.1 Research Setting

This study was conducted at a public university with very high research activity in the Midwestern United States during the Winter 2023 semester. The Institutional Review Board for the Social and Behavioral Sciences of a university in the mid-Atlantic region of the United States approved the study under Protocol #5937. Informed consent was obtained from all participants whose data were analyzed. At the beginning of the semester, 314 students from the Organic Chemistry II course consented to participate in this longitudinal study. Students received course credit for completing the assignments, with 80% awarded for task completion and 20% for the correctness of some closed-response questions. The closed-response questions are not further examined in this study. In total, 266 students completed all four assignments, resulting in 9221 written responses. Data from students who did not complete all assignments were excluded from further analysis.

Of the 266 students, 19 opted not to provide demographic information. Of the remaining 247 participants, 148 identified as a woman, 93 as a man, and six as other gender identities. The students majored in various science-related fields; they were, on average, in their second year of undergraduate studies; and their ages ranged from 18 to 24 years.

3.4.2 Research Instrument

We designed and implemented four online scaffolds on the Qualtrics survey platform to support students' mechanistic reasoning. The scaffolds prompted students to judge the plausibility of alternative mechanisms for a Williamson ether synthesis, enamine hydrolysis, aldol condensation, and Claisen condensation (Fig. 3.3) (Lieber et al., 2022a, b). Each of the four chemical reactions included two plausible and two implausible mechanisms. Students were sequentially presented with the corresponding alternative reaction products associated with these mechanisms but had to identify the underlying mechanisms that lead to these products independently.

Each scaffold consisted of three parts: closed-response tasks to guide students in structuring arguments within domain-general and chemistry-specific scenarios, closed-response tasks to activate students' chemical concept knowledge, and open-ended tasks to prompt students' reasoning about a mechanism's plausibility. The first scaffold served as a baseline for evaluating students' argumentation skills and chemical concept knowledge. Students' open-ended responses on a mechanism's plausibility were automatically scored using a machine learning (ML) algorithm—precisely a transformer-based deep neural network for sequence classification leveraging a pre-trained large language model for natural language processing (Martin et al., 2024). Before conducting this study, a text corpus of 3430 student-written mechanistic reasoning statements was human-coded with an inductively developed rubric (Sect. 3.4.3) and served as training data for the ML algorithm (Martin & Graulich, 2024). The ML algorithm then identified the underlying patterns between students' open-ended reasoning and the human-assigned codes in the training corpus and automatically scored the data gathered within this study. Thus, the ML algorithm replaced human coding by taking students' open-ended reasoning as input and generating a score representing the complexity of their mechanistic reasoning as output (Sect. 3.4.3). We then

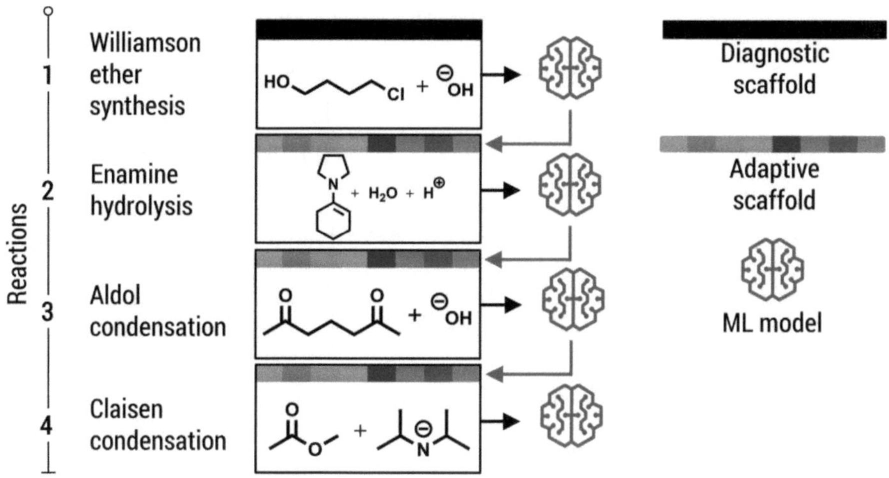

Fig. 3.3 Applying machine learning to adaptively support Organic Chemistry II students' mechanistic reasoning over one semester

averaged students' complexity scores generated by the ML algorithm within each time point of data collection and manually assigned individualized feedback and a tailored set of exercises in Qualtrics to each student aligned with the complexity level demonstrated in their previous responses. This approach allowed us to individualize the second scaffold. After completing the exercises related to enamine hydrolysis, the ML algorithm automatically scored students' open-ended answers again so that subsequent scaffolds could be continually individualized (Fig. 3.3). Our ML analysis has the potential for fine-grained diagnostic analysis of students' mechanistic reasoning (Martin & Graulich, 2023). Besides, our ML-based scoring relates to further projects presented in this book, which include ML applications to quantitatively analyze students' mechanistic reasoning (Watts et al., Chap. 1) along with alternative strategies for leveraging scaffolding and ML to deliver feedback and characterize students' mechanistic reasoning (Cooper et al., Chap. 6).

To scaffold students' reasoning about alternative mechanisms, we used a simplified version of Toulmin's argumentation pattern: the claim-evidence-reasoning model (McNeill & Krajcik, 2011; McNeill et al., 2006). A claim serves as a statement and requires causal support. Causal support comes from explanations or scientific data, known as evidence. This evidence draws on information from various sources to substantiate the claim. The reasoning establishes the cause-effect relation between the claim and evidence. In our setting, students claimed whether the displayed product is (im-)plausible, provided evidence explaining why the product might be (im-)plausible, and connected the claim and evidence through reasoning statements by incorporating electronic, steric, or energetic effects (Fig. 3.4). Sample solutions and student examples can be found in Table 3.3 and elsewhere (Lieber et al., 2022a, b; Martin et al., 2024). Students were given a week to complete each

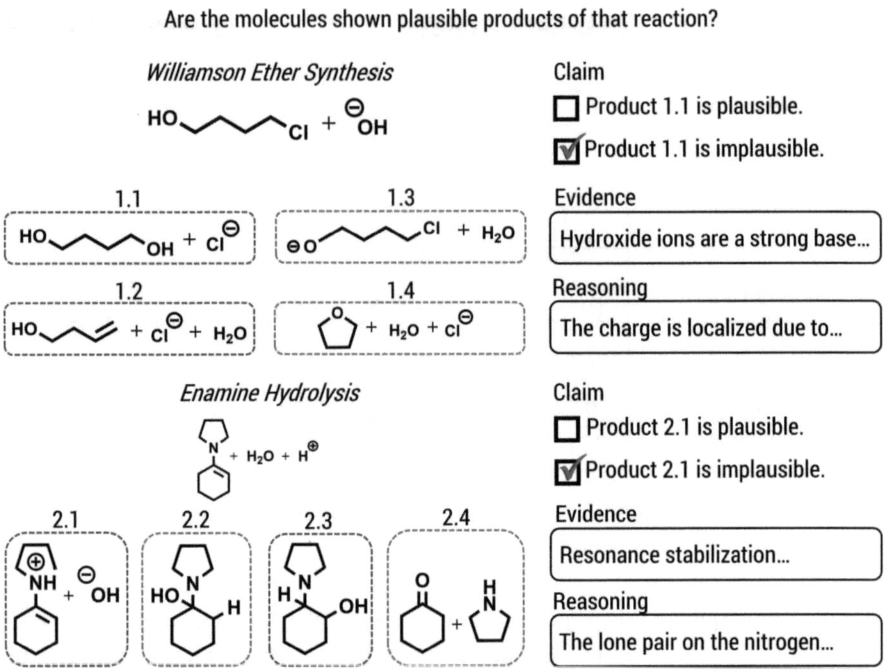

Fig. 3.4 Students judged the plausibility of four alternative mechanisms for four reactions, among others, a Williamson ether synthesis and an enamine hydrolysis. In doing so, students made a claim about the plausibility of the mechanism, justified the claim with evidence, and linked the claim and evidence based on reasoning. Implausible products (1.1, 1.2, 2.1, 2.3) are highlighted in red, and plausible ones in green (1.3, 1.4, 2.2, 2.4)

scaffold. Completing a scaffold took 75–90 min. The scaffolds were scheduled in weeks 6, 8, 12, and 13 of the semester.

3.4.3 Rubric for Machine Scoring

To quantify students' mechanistic reasoning complexity—hereafter simply referred to as *reasoning complexity*—we developed a two-dimensional holistic rubric drawing on the *levels of granularity* (Bodé et al., 2019; Deng & Flynn, 2021) and *modes of reasoning* (Sevian & Talanquer, 2014) (Sect. 3.2). This rubric categorized every response holistically into one reasoning pattern. Based on the applied *level of granularity* (*absent* = 0, *phenomenological* = 1, *electronic* = 2) and *mode of reasoning* (*absent* = 0, *descriptive* = 1, *relational* = 2, *linear causal* = 3), every student-written response was machine-scored for increasing complexity between zero and five (Fig. 3.5). The disciplinary core ideas did not affect this scoring; instead, students' mechanistic reasoning score was solely determined by whether their responses reflected *phenomenological* or *electronic granularity* along with the applied *mode of reasoning*.

		Structural						Energetic	
Disciplinary Core Ideas		Charged Species and Substrates		Nucleophiles and Electrophiles		Acids and Bases		Thermo-dynamics and Kinetics	
Levels of Granularity		Phen.	Elec.	Phen.	Elec.	Phen.	Elec.	Phen.	Elec.
Modes of Reasoning	Absent	0							
	Descriptive	2	3	2	3	2	3	2	3
	Relational	3	4	3	4	3	4	3	4
	Linear Causal	4	5	4	5	4	5	4	5

Fig. 3.5 Rubric to classify students' reasoning complexity. Numbers represent scores students received for their mechanistic reasoning. *Note:* Phen. stands for *phenomenological*; Elec. for *electronic*. Martin et al. (2024) provided a detailed account of the rubric development process

3.5 Methods

We constructed hierarchical linear models (HLMs) with random effects to address the research question. HLMs are statistical models that account for data nested within multiple levels and allow for a flexible analysis of hierarchically structured repeated-measures data (Hox et al., 2017; Raudenbush & Bryk, 2002). Put differently, HLMs extend linear regression by incorporating multiple levels of data—so-called nested data structures—to capture individual and group-level influences on an outcome. In our case, students evaluated the plausibility of alternative mechanisms at four distinct time points during a semester. The analysis is performed at the response level, where the 9221 student-written responses (level 1) are nested within the 266 students (level 2). We specified a two-level model with responses nested within students and time as a varying predictor due to the longitudinal nature of the data.

Each student response is classified according to three categorical predictor variables: the reaction to which it belonged (dummy-coded using the orthogonal contrasts defined in Table 3.2), the actual plausibility or implausibility of the given mechanism (implausible mechanism = 0, plausible mechanism = 1), and a student's claim regarding a mechanism's plausibility (student claim: implausible = 0, student claim: plausible = 1). For example, the students reasoned about mechanism 1.1 (Fig. 3.4), an implausible mechanism that belongs to Williamson ether synthesis; each student had to judge the mechanism's plausibility independently. We investigated the impact of these three response characteristics on students' reasoning complexity classified by the rubric discussed in Sects. 3.2 and 3.4.3.

Normalization or standardization was not required because of the categorical nature of the predictor variables. The maximum-likelihood estimator was used due to the large sample size. We performed the statistical analysis in R (version 4.3.2)

(R Core Team, 2023) with RStudio as an Integrated Development Environment (RStudio Team, 2023) and used the package *nlme* for multilevel modeling (Pinheiro et al., 2011).

3.5.1 Step-by-Step Analysis with Hierarchical Linear Models

Analyzing HLMs involves several steps: assessing the need for an HLM, adding fixed effects, introducing random slopes, and adding interaction terms (Field et al., 2012; Hox et al., 2017; Raudenbush & Bryk, 2002). To evaluate the necessity of an HLM, we first fitted a baseline model that included only an intercept and subsequently allowed intercepts to vary across students, i.e., a random intercept model (Table 3.1: ID 1 and 2). We then compared the two model fits using the Akaike information criterion (AIC) and the Bayesian information criterion (BIC)—statistical metrics used to assess the quality of statistical models. Lower values for these criteria indicate better-fitting models, meaning these models capture the underlying data patterns more accurately while minimizing statistical error. We also performed a likelihood ratio test that compares the deviance—a measure of goodness of fit—of a simpler model to that of a more complex one based on a χ^2-distribution. In our analysis, the more complex model fitted the data better (Table 3.1: ID 2). The adjusted intraclass correlation (ICC) was 0.17 indicating that 17% of the variation in reasoning complexity is due to differences between students, which suggests that an HLM is appropriate in this context.

We hypothesized that students' reasoning complexity differed for mechanisms of varying plausibility, so we added the three response characteristics described above as fixed effects to the random-intercept model. Adding these characteristics as fixed effects significantly improved model fit (Table 3.1: IDs 3, 4, and 5).

Afterward, we tested whether adding a random slope further improved model fit. As typically done for longitudinal data, we allowed the effect of the reaction, i.e., the time point of data collection, to vary across students (Field et al., 2012). A comparison of AIC and BIC values demonstrated a better fit for the random-slopes model (Table 3.1: ID 6). For this reason, the complexity of students' mechanistic reasoning followed different progressions across the semester.

Subsequently, we added interaction terms between the response characteristics to the random-slopes model. Most importantly, the interaction between all three response characteristics was statistically significant (Table 3.1: ID 10), implying that students' reasoning complexity depended on the interplay of all response characteristics. We interpret this best-fitting final model (pseudo-R^2 = 0.18 according to Raudenbush and Bryk (2002)) in Sect. 3.6 using orthogonal contrasts (Table 3.2). For example, contrast C1 compares the Williamson ether synthesis with the combined effects of the other three mechanisms. Additionally, we report Cohen's f^2 as an effect size, with values greater than 0.02 indicating small effects, values greater than 0.15 indicating medium effects, and values greater than 0.35 indicating large effects (Cohen, 1988).

Table 3.1 Model fit statistics

ID	Model	*df*	AIC	BIC	Deviance	Test	*L.Ratio*	*p*-value
Assessing the Need for an HLM								
1	Intercept-Only	2	27,973	27,987	27,969			
2	Random Intercept	3	26,860	26,881	26,854	1 vs 2	1115	<0.001
Adding Fixed Effects								
3	Context	6	26,732	26,774	26,720	2 vs 3	134	<0.001
4	Claim	7	26,687	26,737	26,673	3 vs 4	47	<0.001
5	Plausibility	8	26,674	26,731	26,658	4 vs 5	15	0.001
Introducing Random Slopes								
6	Random Slope	17	26,538	26,659	26,504	5 vs 6	155	<0.001
Adding Two-Way Interactions								
7	Context × Claim	20	26,537	26,680	26,497	6 vs 7	6	0.106
8	Context × Plausibility	23	26,506	26,670	26,460	7 vs 8	37	<0.001
9	Claim × Plausibility	24	26,502	26,673	26,454	8 vs 9	6	0.016
Adding Three-Way Interactions								
10	Context × Claim × Plausibility	27	26,478	26,671	26,424	9 vs 10	30	<0.001

Note: Models build upon each other; for instance, Model 4 includes all effects of Model 3 with *Claim* as an additional fixed effect. *df* degrees of freedom, *AIC* Akaike information criterion, *BIC* Bayesian information criterion, *L.Ratio* likelihood ratio

Table 3.2 Orthogonal contrasts used for dummy-coding the categorical variable *Context* to estimate and report the models

Variable	Contrast	Reference group	Target group
Context	C1	Ether Synthesis	Enamine Hydrolysis + Aldol Condensation + Claisen Condensation
Context	C2	Enamine Hydrolysis	Aldol Condensation + Claisen Condensation
Context	C3	Aldol Condensation	Claisen Condensation

Random effects of the final model suggest that the relationship between response characteristics and reasoning complexity differed significantly in intercepts across students, $SD = 0.45$, 95% CI [0.38, 0.49], $\chi^2(1) = 1115.37$, $p < 0.001$. Similarly, slopes varied across students, $SD = 0.13$, 95% CI [0.10, 0.16], $\chi^2(9) = 154.68$, $p < 0.001$. Slopes and intercepts were negatively correlated, $r = -0.40$, 95% CI [−0.58, −0.18]. The responses' reasoning complexity showed high variability, $SD = 0.96$.

3.5.2 Assumptions Check

We checked whether our model met the necessary assumptions (Hox et al., 2017). Therefore, we initially examined the assumptions related to the least-square residuals resulting from a linear regression separated by student. Visual inspection of a histogram confirmed the normal distribution of these residuals. However, a scatter-plot revealed that homogeneity of variance was not achieved. Moreover, there were some outliers, which is expected given the 9221 student responses. As these outliers were distributed evenly across students, they were not further investigated.

The residuals of the random intercepts and slopes followed a normal distribution as indicated by Shapiro-Wilk tests ($p = 0.248$ for random intercepts, $p = 0.058$ for random slopes). A few outliers were observed for random intercepts and slopes; however, these outliers only represented students with particularly high or low reasoning complexity.

We also conducted a Rainbow test ($F(4611, 4604) = 0.80$, $p = 0.999$), which confirms that the data can be modeled linearly due to the insignificant test statistic. Variance inflation factors ranged from 1.016 to 1.363, indicating the data is not multicollinear. Finally, we examined studentized residuals, Cook's distance, hat values, and the covariance ratio, and found no irregularities.

As not all assumptions were met, we also performed a robust HLM using the *robustlmm* package (Koller, 2016). The results of the robust HLM were consistent with those of the traditional HLM; hence, further details are not discussed.

3.6 Results

Figure 3.6 presents the distribution of students' responses across combinations of the *levels of granularity* and *modes of reasoning*. This distribution reveals that students within the same cohort exhibited diverse mechanistic reasoning patterns with varying complexity, highlighting substantial heterogeneity in how students reason about mechanisms. Moreover, the applied *levels of granularity* and *modes of*

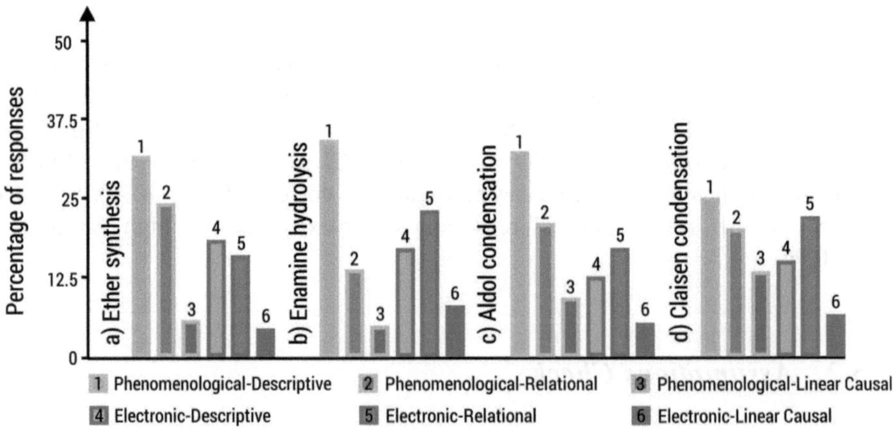

Fig. 3.6 Distribution of students' mechanistic reasoning across combinations of the *levels of granularity* and *modes of reasoning* separated by mechanisms. *Note:* Responses coded as *absent* are excluded from this figure

Table 3.3 Examples of students' mechanistic reasoning on the plausibility of product 1.4 in the ether synthesis by *levels of granularity* and *modes of reasoning*

Level of granularity	Mode of reasoning	Student example
Phenomenological	Descriptive	The hydroxide ion deprotonates the alcohol. The acid-base reaction will proceed due to the strong base.
Phenomenological	Relational	The hydroxide ion can remove the acidic proton of the alcohol. The pK_a values of H_2O and a primary alcohol are about the same, so the hydroxyl may be deprotonated.
Phenomenological	Linear Causal	The hydroxide ion deprotonates the alcohol. The conjugate acid of the ^-OH has a pK_a of 15.7 and the alcohol has a pK_a of 16. The ^-OH will act as a base. Protons can be exchanged in equilibrium since the pK_a-values are so close.
Electronic	Descriptive	The ^-OH is likely to deprotonate the alcohol to become more stable. An ^-OH group is not stable due to the negative charge.
Electronic	Relational	The oxygen of the alcohol becomes negatively charged after the ^-OH attacks (…). The negatively charged oxygen of the intermediate is electron-rich due to electron-donating effects, so it attacks the carbon. This causes the chlorine to leave, resulting in a more stable structure based on molecular principles.
Electronic	Linear Causal	This reaction starts with the deprotonation of the hydroxyl group of the primary alcohol. The hydroxyl group is the most acidic site. Its deprotonation leads to a localized negative charge on the oxygen due to the high electronegativity of the oxygen and the electron-donating effect of the alkyl group. Finally, intramolecular substitution results in the ejection of the chloride ion and the formation of a low-strain ring structure.

reasoning differed across contexts, implying that students may face context-dependent challenges in demonstrating mechanistic reasoning. Table 3.3 presents examples of students' mechanistic reasoning about the plausibility of product 1.4 in the ether synthesis (Fig. 3.4), categorized by the six combinations of the *levels of granularity* and *modes of reasoning*.

To investigate how students' reasoning complexity differed among mechanisms of varying plausibility, we interpreted the best-fitting HLM (Table 3.1: ID 10). This model includes a significant three-way interaction, implying that a multifaceted interplay of response characteristics impacts students' reasoning complexity (Table 3.4). Due to the significant three-way interaction, main effects and lower-level interactions should be interpreted cautiously.

First, the context, i.e., the reaction that a response addressed, significantly predicted students' reasoning complexity (Table 3.4: ID 10.2). Students' mechanistic reasoning was less complex in the ether synthesis compared to the combination of all other reactions (Table 3.4: ID 10.2a), comparable in the enamine hydrolysis and condensation reactions (Table 3.4: ID 10.2b), and more complex in Claisen condensation than in the aldol condensation (Table 3.4: ID 10.2c). These differences show that students' reasoning complexity differed across the four reactions

(Fig. 3.7a). Furthermore, students' individual claim regarding the plausibility of a mechanism also predicted their reasoning complexity (Table 3.4: ID 10.3). Here, the reasoning complexity was higher when students argued *for* the plausibility of a mechanism (Fig. 3.7b). Conversely, the actual plausibility of a mechanism did not significantly influence students' reasoning complexity (Table 3.4: ID 10.4; Fig. 3.7c).

Moreover, we investigated the two-way interactions, which assess the combined effects of two of the three response characteristics on students' reasoning complexity. Here, the context and students' individual claims are related (Table 3.4: ID 10.5), implying that the increased reasoning complexity when arguing *for* the plausibility of a mechanism depended significantly on the reaction. Precisely, students' reasoning complexity was higher—in part insignificantly—when arguing *for*, rather than *against* a mechanism's plausibility for enamine hydrolysis, aldol condensation, and Claisen condensation; this difference was even more pronounced for Williamson ether synthesis (Fig. 3.8a). Orthogonal contrasts confirm that variations in students' reasoning complexity considering their claims were more pronounced for Williamson ether synthesis compared to the other reactions (Table 3.4: ID 10.5a), whereas variations were similar among enamine hydrolysis, aldol condensation, and Claisen condensation (Table 3.4: IDs 10.5b and 10.5c). Similarly, the context and a mechanism's actual plausibility are related (Table 3.4: ID 10.6), indicating that students' reasoning complexity for plausible and implausible mechanisms differed significantly across reactions. Students' reasoning complexity was comparable for plausible and implausible mechanisms in the aldol condensation and Claisen condensation, higher for plausible mechanisms in Williamson ether synthesis, but significantly lower for plausible mechanisms in enamine hydrolysis (Fig. 3.8b). Orthogonal contrasts confirm that variations in students' reasoning complexity regarding a mechanism's actual plausibility were more pronounced for the ether synthesis compared to the combined effect of the other mechanisms (Table 3.4: ID 10.6a). Furthermore, students demonstrated significantly more complex mechanistic reasoning about implausible mechanisms in enamine hydrolysis compared to the condensation reactions (Table 3.4: IDs 10.6b); variations in reasoning complexity between plausible and implausible mechanisms were comparable for the aldol and Claisen condensations (Table 3.4: ID 10.6c). Beyond that, a student's claims and a mechanism's actual plausibility are also related (Table 3.4: ID 10.7). This interaction shows that the reasoning complexity increased when students' individual claims about a mechanism's plausibility aligned with its actual plausibility, which indicates that students' mechanistic reasoning was more complex when their claim was correct (Fig. 3.8c).

Most importantly, the interaction between all three response characteristics is also statistically significant (Table 3.4: ID 10.8). This three-way interaction reveals that a correct claim did not equally increase students' reasoning complexity across all reactions. For example, students' mechanistic reasoning was more complex with correct claims in enamine hydrolysis, aldol condensation, and Claisen condensation (Fig. 3.9b, c, and d). By contrast, students' reasoning was more complex when they

Table 3.4 Parameters of the best-fitting final model (Table 3.1: ID 10)

				95% *CI*				
ID	Fixed effects	Contrast	*b*	*LL*	*UL*	*t*-value	*p*-value	Cohen's f^2
10.1	Intercept	–	2.883	2.817	2.949	85.417	<0.001	0.816
10.2a	Context	C1	0.092	0.066	0.118	6.944	<0.001	0.005
10.2b	Context	C2	−0.025	−0.059	0.010	−1.570	0.117	<0.001
10.2c	Context	C3	0.154	0.105	0.203	6.155	<0.001	0.004
10.3	Claim	–	0.121	0.043	0.199	3.045	0.002	0.001
10.4	Plausibility	–	0.016	−0.101	0.133	0.268	0.789	<0.001
10.5a	Context × Claim	C1	−0.086	−0.124	−0.049	−4.507	<0.001	0.002
10.5b	Context × Claim	C2	−0.023	−0.086	0.039	−0.733	0.464	<0.001
10.5c	Context × Claim	C3	0.043	−0.081	0.167	0.683	0.494	<0.001
10.6a	Context × Plausibility	C1	−0.155	−0.198	−0.111	−6.987	<0.001	0.006
10.6b	Context × Plausibility	C2	−0.038	−0.072	−0.003	−2.122	0.034	0.001
10.6c	Context × Plausibility	C3	0.107	−0.039	0.253	1.431	0.152	<0.001
10.7	Claim × Plausibility	–	0.248	0.127	0.368	4.032	<0.001	0.002
10.8a	Context × Claim × Plausibility	C1	0.155	0.098	0.211	5.348	<0.001	0.003
10.8b	Context × Claim × Plausibility	C2	0.031	−0.064	0.126	0.642	0.521	<0.001
10.8c	Context × Claim × Plausibility	C3	−0.138	−0.331	0.056	−1.395	0.163	<0.001

Note: The definitions of contrasts C1-C3 are provided in Table 3.2. *Degrees of freedom* 8940, *b* regression coefficient, *LL* lower limit, *UL* upper limit

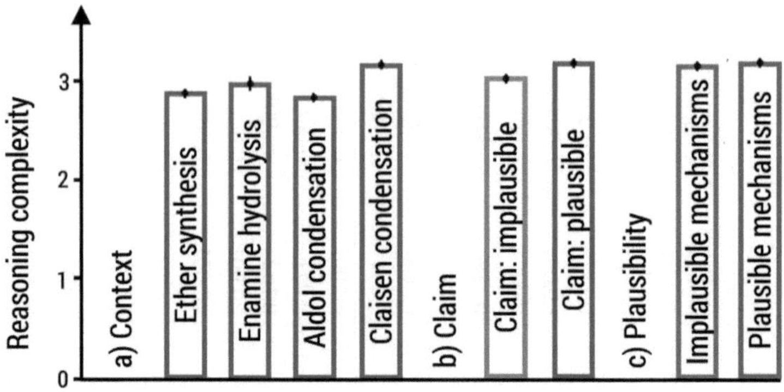

Fig. 3.7 Error bar charts visualizing how the context, students' claims, and a mechanism's actual plausibility impact students' reasoning complexity

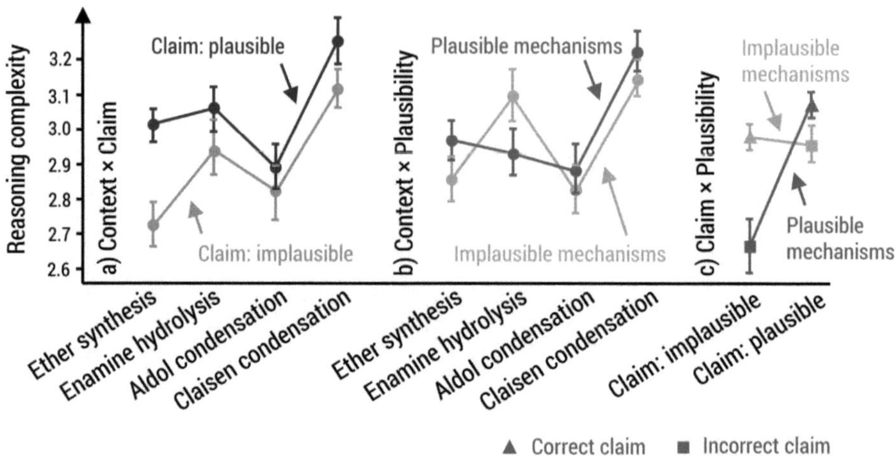

Fig. 3.8 Line graphs visualizing how two of three response characteristics impact students' reasoning complexity. *Note:* Blue indicates that students claim a mechanism's plausibility and orange that they claim a mechanism's implausibility; green symbolizes plausible mechanisms and red implausible ones; triangles visualize correct claims and squares incorrect claims

argued *for* the plausibility of a mechanism in the ether synthesis, irrespective of a mechanism's actual plausibility (Fig. 3.9a). Orthogonal contrasts confirm that the trend of students having more complex reasoning when their claims were correct does not hold for Williamson ether synthesis (Table 3.4: IDs 10.8a) but applies to the other three mechanisms (Table 3.4: IDs 10.8b and 10.8c). Overall, this result suggests that choosing the correct claim is typically the key response characteristic causing complex mechanistic reasoning in most mechanisms—except for the ether synthesis.

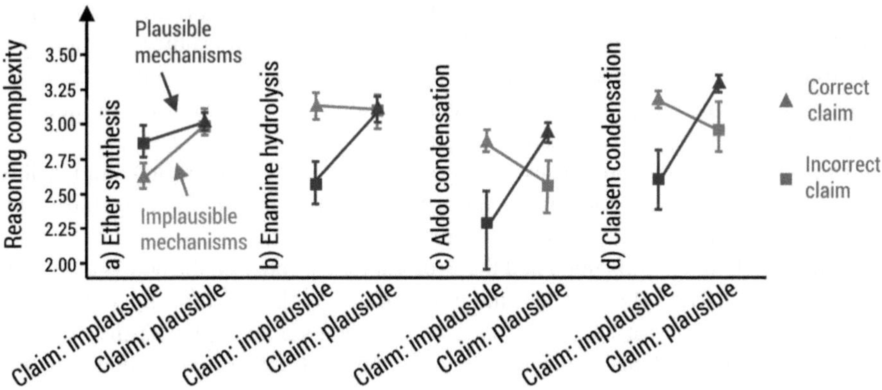

Fig. 3.9 Line graphs visualizing how the three response characteristics impact students' reasoning complexity. *Note:* Green symbolizes plausible mechanisms and red implausible ones; triangles visualize correct claims and squares incorrect claims

3.7 Discussion

Our study revealed how the complexity of students' mechanistic reasoning classified by the first and third essential epistemic heuristics varies across different response characteristics when evaluating a mechanism's plausibility. The analysis showed that students' reasoning complexity ranged roughly from 2.25 to 3.25, indicating that the responses examined are mostly constructed at a *descriptive* (44%) or *relational* (35%) *mode of reasoning* and a *phenomenological* (56%) *level of granularity* (Fig. 3.6). This finding aligns with our prior analyses classifying students' reasoning on alternative mechanisms in different settings (Martin et al., 2024). The lack of causal connections in students' mechanistic reasoning is a common observation across disciplines (Flowers et al., 2023). Nevertheless, students' reasoning complexity differed in our setting significantly across the three examined response characteristics, which implies that certain characteristics promote complex reasoning about a mechanism's plausibility.

First, our findings demonstrate that students' mechanistic reasoning is more complex when arguing *for*, rather than *against* a mechanism's plausibility (Fig. 3.7b). This finding can be rationalized with common instructional practices in organic chemistry that emphasize only the most plausible mechanism as determined by the lowest energy path (Caspari et al., 2018; Goodwin, 2012). As a result, students are more often prompted to argue *for*, rather than *against* a given outcome, so they tend to concentrate only on the most plausible products, neglecting potential byproducts (Lieber & Graulich, 2022; Lieber et al., 2022b; Popova & Bretz, 2018). This common practice can lead to a lower reasoning complexity when arguing *against* a mechanism's plausibility. The observed difference in reasoning complexity was even more pronounced in the Williamson ether synthesis (Fig. 3.8a). This reaction was the first context in our study where students encountered implausible mechanisms, so one can assume that students were less familiar with justifying a mechanism's implausibility, which decreased students' reasoning complexity. Continually practicing to argue not only *for* but also *against* potential mechanisms may help students recognize that chemical reactions can yield multiple products of different plausibility, unpack the key factors influencing a mechanism's plausibility, and link the emerging properties to support their claims. Integrating implausible alternatives and contrary viewpoints more explicitly into instruction may, thus, support students' mechanistic reasoning (Larrain et al., 2019).

While students' claims impacted their reasoning complexity, the actual plausibility of a mechanism did not significantly influence their reasoning complexity (Fig. 3.7c). This is not surprising as a mechanism's actual plausibility was not accessible to the students in our setting since they did not receive feedback on the correctness of their claims. Surprisingly, students' mechanistic reasoning was more complex for implausible mechanisms in enamine hydrolysis while this trend was reversed for the other three mechanisms (Fig. 3.8b). The implausible mechanisms in enamine hydrolysis arose from an acid-base reaction and an electrophilic addition in the anti-Markovnikov position, i.e., a typically used rule for asymmetric alkene addition, which were probably clearly implausible to the students. Future analyses need to investigate this difference for enamine hydrolysis in detail.

As further shown, students' reasoning complexity was most importantly influenced by the correctness of their claims, with correct claims leading to more complex mechanistic reasoning (Fig. 3.8c). This finding corresponds to other qualitative and quantitative analyses: For example, Cruz-Ramírez de Arellano and Towns (2014) found that students with correct claims about alkyl halide reactions could more thoroughly weigh basic and nucleophilic properties of substances, accurately argue about intermediates and transition states, and build chemically sound reasoning statements to connect entities, activities, and their properties. Also, students' *modes of reasoning* were significantly higher for students with correct claims than for those with incorrect claims when judging which of three bases can best force the acid-base equilibrium to the product side (Deng & Flynn, 2021). By contrast, a similar analysis showed that the proportion of the *modes of reasoning* did not vary significantly between responses with correct and incorrect claims when contrasting the plausibility of two unimolecular nucleophilic substitutions (Bodé et al., 2019). Our analysis builds on this work by demonstrating that a correct claim may lead to a higher reasoning complexity (Fig. 3.8c); however, this trend does not apply to the Williamson ether synthesis (Fig. 3.9). The reactions evaluated by Bodé et al. (2019) and the Williamson ether synthesis analyzed herein both involve nucleophilic substitution (S_N) mechanisms, so unsurprisingly, these reactions deviate from the observed trend. Students' familiarity with the S_N mechanisms—which are introduced early in undergraduate organic chemistry and follow a clear reaction pattern—may reduce the role of claim-making in students' decision-making. Instead of engaging in mechanistic reasoning, students likely relied on their existing knowledge and recalled the S_N portions of the reactions (Grove et al., 2012). This may have allowed the students to arrive at conclusions more confidently without a time-intensive, complex decision-making process. Nonetheless, the significant impact of making a correct claim on students' reasoning complexity could encourage instruction to guide the process of constructing and justifying claims more consciously (McNeill & Pimentel, 2010).

Ultimately, students' reasoning complexity differed significantly across reactions and tended to increase across the semester (Fig. 3.7a). This increase was expected due to continuous adaptive support in our setting that helped students use chemical concepts to reason causally. A previous comparable adaptive training also provided students effective support in building scientific arguments and applying chemical concepts (Lieber et al., 2022a, b), so it is unsurprising that students' reasoning complexity tends to increase over time. However, aldol condensation seems particularly challenging for students due to the complex interplay of acid-base equilibria, carbonyl chemistry, and elimination reactions (Watts et al., 2022). This context-dependency of students' mechanistic reasoning transcends disciplinary boundaries as exemplified by an analysis of students' mechanistic reasoning in molecular and cellular biology (Southard et al., 2017).

3.8 Conclusions and Implications

To investigate students' reasoning complexity when judging the plausibility of alternative mechanisms, we operationalized the *essential epistemic heuristics for mechanistic reasoning* (Krist et al., 2019). Specifically, we compared the complexity of students' mechanistic reasoning related to granularity and causality—reflecting the first and third essential epistemic heuristics—to identify the response characteristics that support complex mechanistic reasoning about alternative mechanisms. Our findings hold various implications for guiding undergraduate students' mechanistic reasoning in science classes.

First, our analysis revealed that students are better at arguing *for* than *against* the plausibility of a mechanism. To support students' mechanistic reasoning in science classes, instruction can progressively increase the difficulty of exercises by initially presenting students with plausible mechanisms and gradually incorporating implausible ones throughout the semester. In this context, discussions about mechanisms of different plausibility should occur explicitly. For instance, when discussing competitive mechanisms, such as substitution and elimination reactions, instruction should highlight that there can be multiple, sometimes conflicting interpretations of a mechanism's outcome, rather than viewing this outcome as strictly right or wrong with little room for ambiguity (Grove & Bretz, 2010). Advancing students' ability to critically evaluate conflicting information, for example, mechanistic hypotheses, may finally enrich the complexity of students' reasoning in everyday life.

Furthermore, students demonstrated greater complexity in reasoning across scalar levels and linking properties at lower scalar levels if their claims were correct. This finding suggests that correct claim-making is crucial for the ability to transition between scalar levels and establish causal links, which means correct claim-making and mechanistic reasoning benefit each other (Choi et al., 2013; Grove et al., 2012; Lieber et al., 2022a; Walker et al., 2019). Consequently, instruction should explicitly support students' claim-making by introducing plausible *and* implausible mechanisms.

In sum, our findings may inform the development of a step-by-step instructional sequence that guides students in reasoning about multiple mechanistic alternatives. From a broader perspective, encouraging students to consider alternative hypotheses or explanations can help them evaluate evidence, weigh different possibilities, and make informed judgments about scientific phenomena (Driver et al., 2000). Such guidance can support students in looking beyond a simplistic view of mechanisms as a linear sequence of events; instead, it can encourage them to reason about the dynamic interplay of entities and activities by weighing various alternatives.

3.9 Limitations

To identify response characteristics that support complex mechanistic reasoning, we examined how the complexity of students' mechanistic reasoning differed across mechanisms of varying plausibility. Our analysis focused on the response level. Further insights into students' mechanistic reasoning could be provided by centering our analysis at the student level. Examining each student, rather than every response, may provide a more holistic view of students' reasoning complexity. This allows, on the one hand, to model students' mechanistic reasoning paths across the four reactions. On the other hand, we may identify the cognitive and affective variables that favor complex mechanistic reasoning. Accordingly, future analyses could adopt a student-centered perspective to uncover productive mechanistic reasoning paths and explore their causes.

References

Bhattacharyya, G., & Bodner, G. M. (2005). "It gets me to the product": How students propose organic mechanisms. *Journal of Chemical Education, 82*(9), 1402–1407.

Bodé, N. E., Deng, J. M., & Flynn, A. B. (2019). Getting past the rules and to the WHY: Causal mechanistic arguments when judging the plausibility of organic reaction mechanisms. *Journal of Chemical Education, 96*(6), 1068–1082.

Caspari, I., Weinrich, M., Sevian, H., & Graulich, N. (2018). This mechanistic step is "*productive*": Organic chemistry students' backward-oriented reasoning. *Chemistry Education Research and Practice, 19*(1), 42–59.

Choi, A., Hand, B., & Greenbowe, T. (2013). Students' written arguments in general chemistry laboratory investigations. *Research in Science Education, 43*(5), 1763–1783.

Cohen, J. (1988). *Statistical power analysis for the behavioral sciences* (Vol. 2). Erlbaum.

Craver, C. F. (2007). *Explaining the brain: Mechanisms and the mosaic unity of neuroscience* (Vol. 1). Clarendon Press.

Cruz-Ramírez de Arellano, D., & Towns, M. H. (2014). Students' understanding of alkyl halide reactions in undergraduate organic chemistry. *Chemistry Education Research and Practice, 15*(4), 501–515.

Darden, L., & Craver, C. (2002). Strategies in the interfield discovery of the mechanism of protein synthesis. *Studies in History and Philosophy of Biological and Biomedical Sciences, 33*(1), 1–28.

Deng, J. M., & Flynn, A. B. (2021). Reasoning, granularity, and comparisons in students' arguments on two organic chemistry items. *Chemistry Education Research and Practice, 22*(3), 749–771.

Driver, R., Newton, P., & Osborne, J. (2000). Establishing the norms of scientific argumentation in classrooms. *Science Education, 84*(3), 287–312.

Field, A., Miles, J., & Field, Z. (2012). *Discovering statistics using R* (Vol. 1). Sage.

Flowers, S., Holder, K. H., & Gardner, S. M. (2023). Here is the biology, now what is the mechanism? Investigating biology undergraduates' mechanistic reasoning within the context of biofilm development. *Journal of Microbiology & Biology Education, 24*(2), e00221–00222.

Goodwin, W. M. (2003). Explanation in organic chemistry. *Annals of the New York Academy of Sciences, 988*(1), 141–153.

Goodwin, W. M. (2012). Mechanisms and chemical reaction. In A. I. Woody, R. F. Hendry, & P. Needham (Eds.), *Philosophy of chemistry* (Vol. 6, pp. 309–327). North-Holland.

Grove, N. P., & Bretz, S. L. (2010). Perry's scheme of intellectual and epistemological development as a framework for describing student difficulties in learning organic chemistry. *Chemistry Education Research and Practice, 11*(3), 207–211.

Grove, N. P., Cooper, M. M., & Cox, E. L. (2012). Does mechanistic thinking improve student success in organic chemistry? *Journal of Chemical Education, 89*(7), 850–853.

Hox, J., Moerbeek, M., & van de Schoot, R. (2017). *Multilevel analysis: Techniques and applications* (Vol. 3). Routledge.

Illari, P. M., & Williamson, J. (2012). What is a mechanism? Thinking about mechanisms across the sciences. *European Journal for Philosophy of Science, 2*(1), 119–135.

Koller, M. (2016). *robustlmm: Robust linear mixed effects models* [Computer Program].

Krist, C., Schwarz, C. V., & Reiser, B. J. (2019). Identifying essential epistemic heuristics for guiding mechanistic reasoning in science learning. *Journal of the Learning Sciences, 28*(2), 160–205.

Larrain, A., Freire, P., López, P., & Grau, V. (2019). Counter-arguing during curriculum-supported peer interaction facilitates middle-school students' science content knowledge. *Cognition and Instruction, 37*(4), 453–482.

Lieber, L. S., & Graulich, N. (2022). Investigating students' argumentation when judging the plausibility of alternative reaction pathways in organic chemistry. *Chemistry Education Research and Practice, 23*(1), 38–53.

Lieber, L. S., Ibraj, K., Caspari-Gnann, I., & Graulich, N. (2022a). Closing the gap of organic chemistry students' performance with an adaptive scaffold for argumentation patterns. *Chemistry Education Research and Practice, 23*(4), 811–828.

Lieber, L. S., Ibraj, K., Caspari-Gnann, I., & Graulich, N. (2022b). Students' individual needs matter: A training to adaptively address students' argumentation skills in organic chemistry. *Journal of Chemical Education, 99*(7), 2754–2761.

Machamer, P., Darden, L., & Craver, C. F. (2000). Thinking about mechanisms. *Philosophy of Science, 67*(1), 1–25.

Martin, P. P., & Graulich, N. (2023). When a machine detects student reasoning: A review of machine learning-based formative assessment of mechanistic reasoning. *Chemistry Education Research and Practice, 24*(2), 407–427.

Martin, P. P., & Graulich, N. (2024). Navigating the data frontier in science assessment: Advancing data augmentation strategies for machine learning applications with generative artificial intelligence. *Computers and Education: Artificial Intelligence, 7*, 100265.

Martin, P. P., Kranz, D., Wulff, P., & Graulich, N. (2024). Exploring new depths: Applying machine learning for the analysis of student argumentation in chemistry. *Journal of Research in Science Teaching, 61*(8), 1757–1792.

McNeill, K. L., & Krajcik, J. (2011). *Supporting grade 5–8 students in constructing explanations in science: The claim, evidence, and reasoning framework for talk and writing* (Vol. 1). Pearson Allyn & Bacon.

McNeill, K. L., & Pimentel, D. S. (2010). Scientific discourse in three urban classrooms: The role of the teacher in engaging high school students in argumentation. *Science Education, 94*(2), 203–229.

McNeill, K. L., Lizotte, D. J., Krajcik, J., & Marx, R. W. (2006). Supporting students' construction of scientific explanations by fading scaffolds in instructional materials. *Journal of the Learning Sciences, 15*(2), 153–191.

Pinheiro, J., Bates, D., & R Core Team. (2011). *nlme: Linear and nonlinear mixed effects models* [Computer Program].

Popova, M., & Bretz, S. L. (2018). "It's only the major product that we care about in organic chemistry": An analysis of students' annotations of reaction coordinate diagrams. *Journal of Chemical Education, 95*(7), 1086–1093.

R Core Team. (2023). *R: A language and environment for statistical computing* [Computer Program].

Raudenbush, S. W., & Bryk, A. S. (2002). *Hierarchical linear models: Applications and data analysis methods* (Vol. 1). Sage.

RStudio Team. (2023). *RStudio: Integrated development environment for R* [Computer Program].

Russ, R. S., Scherr, R. E., Hammer, D., & Mikeska, J. (2008). Recognizing mechanistic reasoning in student scientific inquiry: A framework for discourse analysis developed from philosophy of science. *Science Education, 92*(3), 499–525.

Sevian, H., & Talanquer, V. (2014). Rethinking chemistry: A learning progression on chemical thinking. *Chemistry Education Research and Practice, 15*(1), 10–23.

Southard, K. M., Espindola, M. R., Zaepfel, S. D., & Bolger, M. S. (2017). Generative mechanistic explanation building in undergraduate molecular and cellular biology. *International Journal of Science Education, 39*(13), 1795–1829.

Talanquer, V. (2013). When atoms want. *Journal of Chemical Education, 90*(11), 1419–1424.

Walker, J. P., Van Duzor, A. G., & Lower, M. A. (2019). Facilitating argumentation in the laboratory: The challenges of claim change and justification by theory. *Journal of Chemical Education, 96*(3), 435–444.

Watts, F. M., Park, G. Y., Petterson, M. N., & Shultz, G. V. (2022). Considering alternative reaction mechanisms: Students' use of multiple representations to reason about mechanisms for a writing-to-learn assignment. *Chemistry Education Research and Practice, 23*(2), 486–507.

Weinrich, M. L., & Sevian, H. (2017). Capturing students' abstraction while solving organic reaction mechanism problems across a semester. *Chemistry Education Research and Practice, 18*(1), 169–190.

Chapter 4
Towards a Wide Understanding of Mechanisms Across Science Domains: Implications for Science Education

Vanessa De Andrade, Yael Shwartz, and Pedro Correia Rodrigues

4.1 Introduction

Mechanistic reasoning has gained the attention of science education researchers. When students engage in this way of thinking, they move beyond acquiring scientific facts; instead, they actively grasp how things work (Andrade et al., 2022; Krist et al., 2019; Tang et al., 2020; see also Chaps 5 and 10). Developing explanations through the discovery of mechanisms provides students with valuable thinking resources for generating, understanding and evaluating explanations across various fields, from science to technology to everyday life. This aligns with the current calls for preparing autonomous students who can guide their own scientific intellectual work (NGSS, 2013; NRC, 2012) and, more generally, calls for educating for thinking (OECD, 2016).

The emphasis on mechanistic reasoning and how it guides students' explanation construction has led to the development of various frameworks (e.g., Andrade et al., 2019; Krist et al., 2019; Russ et al., 2008; van Mil et al., 2013). These frameworks have allowed us to better characterise students' mechanistic reasoning and explanations and gain a deeper understanding of how they progress in their explanatory accounts, the challenges they encounter, and how to guide them towards more sophisticated mechanistic accounts. An early characterization of mechanism in the life sciences by the philosophers of science, Machamer, Darden, and Craver (2000),

V. De Andrade (✉) · P. C. Rodrigues
Centre for Research and Innovation in Education (inED), ESE, Polytechnic of Porto, Porto, Portugal
e-mail: vanessafigueiredo@campus.ul.pt; Vanessa.de-andrade@weizmann.ac.il

Y. Shwartz
Department of Science Teaching, Weizmann Institute of Science, Rehovot, Israel

N. Graulich et al. (eds.), *Exploring Mechanistic Reasoning in Science Education*, Contributions from Science Education Research 17,
https://doi.org/10.1007/978-3-032-10246-1_4

has informed much of these frameworks. Most of these frameworks reflect the epistemic work involved in constructing mechanistic explanations within specific fields, primarily life science and chemistry; however, less has been discussed regarding what we are calling a mechanism, how it characterizes, and how mechanistic reasoning might apply across different science domains.

In this chapter, we examine the recent literature on mechanistic reasoning within the philosophy of science—particularly the work of Illari and Williamson (2012) as well as Glennan, Illari, and Weber (2022). Key aspects of this chapter include offering a minimal definition of mechanisms that synthesises many of the extant ideas about mechanisms. The aim is to show how this minimal definition applies widely across various scientific fields and illustrate its value in guiding mechanism discovery in various scientific phenomena.

4.2 What Is a Mechanism? —A Wide Framework

In this section, we will present the definition of a mechanism guiding our work. We have borrowed a characterization of a mechanism from Illari and Williamson (2012).[1] Illari and Williamson's proposal aims to synthesise nearly two decades of debate in the philosophy of science on how to characterise a mechanism. An influential and early work on this matter was a paper by Machamer, Darden, and Craver (2000). Other researchers followed this work, including Bechtel and Abrahamsen (2005) and Glennan (2002); among others, see also Craver et al. (2024). The work of these authors is associated with what is called "New Mechanism". Originally, New Mechanist philosophers concentrated on the molecular life sciences, for which covering law-type explanations were not suited, as there are hardly any laws of nature in biology (Glennan et al., 2022). The success of mechanism-based explanations in the life sciences has led to the often considered (erroneously) that this type of explanation is peculiar to this field (Glennan et al., 2022). Nevertheless, the centrality of the search for mechanisms has since garnered interest beyond the life sciences, including physics, ecology, environmental issues, social sciences, etc. The thesis that "mechanistic explanations are ubiquitous across the empirical sciences" (Glennan et al., 2022, p. 148) advocates for a widely applicable understanding of mechanisms, one that can be useful in understanding the specific types of mechanisms within a particular domain. To uncover similarities among mechanisms across scientific fields, Illari and Williamson (2012, p. 120) proposed the following minimal definition:

> A mechanism for a phenomenon consists of entities whose activities and interactions are organized so as to be responsible for the phenomenon.

[1] The term "wide" is borrowed from Illari and Williamson (2012) and is used in the sense that this framework draws on a minimal definition of mechanism to capture the commonalities among mechanisms across various fields that can apply broadly.

With this consensus definition of mechanisms, the authors summarise and illuminate important major points of consensus among previous proposals, particularly among three early pivotal definitions of mechanisms and frequently cited:

> Mechanisms are entities and activities organized such that they are productive of regular changes from start or set-up to finish or termination conditions. (Machamer et al., 2000, p. 3)

> A mechanism for a behaviour is a complex system that produces that behaviour by the interaction of a number of parts, where the interactions between parts can be characterised by direct, invariant, change-relating generalisations. (Glennan, 2002, p. S344)

> A mechanism is a structure performing a function in virtue of its component parts, component operations, and their organization. The orchestrated functioning of the mechanism is responsible for one or more phenomena. (Bechtel & Abrahamsen, 2005, p. 423)

Over two decades of discussion on defining mechanisms have resulted in some points of consensus embodied in Illari and Williamson's minimal definition. The authors clarify that their definition does not claim to have discovered the true nature of mechanisms, but instead aims to provide a framework for analysing mechanisms across scientific domains. Our rationale for adopting Illari and Williamson's minimal definition is that it will be useful for guiding reflection on current and future work concerning mechanistic reasoning in science education. For instance, what are we calling a mechanism in the various disciplines? What do the many disciplines have in common that we can use to learn about and guide students' mechanistic reasoning?

For the sake of wide applicability, Illari and Williamson's definition omits some concepts of potential conflict while highlighting what seems consensual among the early proposals. Here, we outline these conflicts and the associated discussions.

4.2.1 Are All Mechanisms "Structures" or "Complex Systems"?

In Bechtel and Abrahamsen's (2005) framework, a "mechanism is a structure," whereas in Glennan's, it is a "complex system." Illari and Williamson (2012), aiming for a broader understanding, avoid terms like "structure" or "complex system", permitting the mechanism to encompass any of these and other possible arrangements. The authors contend that the concept of structure implies a rigidity that may not be applicable to all mechanisms. While systems are generally more dynamic and adaptable than structures, they also demonstrate regularity, stability, and persistence—traits often absent in many mechanisms. For example, one-off processes leading to singular events, such as asteroid collisions or genetic mutations, do not possess these systemic features (Glennan, 2017). Furthermore, not all mechanisms are complex systems; some are relatively simple. Illari and Williamson (2012) illustrate this with the thermal dissociation of diatomic iodine in the vapour phase, where increasing vibrational energy ultimately breaks the bond between atoms. This appears too simplistic to be considered a "complex system." Thus, they argue for a more flexible interpretation that encompasses mechanisms across the continuum from rigid to flexible structures, from regular to on-time events, and from simple to complex systems.

4.2.2 Do All Mechanisms Produce Regular Changes?

Some mechanisms do not bring about any change. For instance, consider homeostatic mechanisms, which work to maintain a state of affairs rather than to bring about a change. Also, some mechanisms work irregularly, and others work only once (Craver & Tabery, 2019). Machamer (2011) provides us with the following example: Supernovae are a general class, but the one in 1604 was unique within its category of phenomena and, thus, its mechanism. Illari and Williamson (2012) contend that instead of claiming that mechanisms do not require some form of regularity or stability, abandoning the concept of regularity seeks to encompass the various forms of stability that mechanisms may exhibit. Furthermore, the authors state that the very concept of *mechanisms responsible for* already embodies some form of lesser regularity or stability.

4.2.3 Do All Mechanisms Have Start (or Step-Up) and Finish (or Terminal) Conditions?

Illari and Williamson (2012) also leave out Machamer et al.'s (2000) proposal that mechanisms work from initial to terminal conditions. They argue that many mechanisms have no actual step-up or terminal conditions. For continuing mechanisms, there are no true starts or endings. Some mechanisms act in a cyclical way, such as the Krebs cycle. Others, like the day-night cycle or the moon's phases, can be described in terms of a starting and an ending, but the mechanism itself responsible for the phenomenon does not possess true start or finish conditions. Bechtel (2009) has similarly noted that how a mechanism operates greatly relies on the surrounding conditions, which also encompasses any regularities within that context. However, these conditions are not explained by the mechanism itself, its parts, and its operations; rather, they are described by examining the environment in which the mechanism works. Thus, Illari and Williamson argue that a wide framework should not enforce the idea of mechanisms working from start to finish conditions.

4.2.4 Do All Mechanisms Perform a Function?

Bechtel and Abrahamsen (2005) propose that mechanisms are structures performing a function. Illari and Williamson refrain from explicitly using the term "performing a function" within a wide framework. They argue that the concept of "function" might lead to a restrictive understanding of mechanisms that implies some form of intentional design or selection, while the "function" of a mechanism might require only certain "characteristic activities" (p. 124). There are mechanisms that are responsible for phenomena, but they are not designed for or do not function to cause them (Glennan, 2017). For example, phenomena in chemistry are the result of emergent mechanisms, i.e., the properties or behavior of a phenomenon result from the dynamic collections of the interacting entities (Talanquer, 2018) that serve no purpose. In astronomy, many phenomena result from a characteristic spatial and temporal organization, where the position of an entity in relation to another and the duration of an entity's activity bring about the phenomenon. These types of mechanisms may not be best characterised as structures that perform a function.

In addressing these potentially problematic cases, Illari and Williamson (2012) aimed to propose a more encompassing definition—one that does not overlook things that are intuitively understood as mechanisms but may be excluded due to potential conflicts. In doing so, they drew on commonalities found in earlier proposals. Therefore, within a wide framework:

> mechanisms are identified and individuated by (a) what they do (the "phenomenon"), (b) what they are made of ("components"/"entities"/"parts"), what their parts do ("activities"/"interactions"/"operations"), and (c) how they are structured ("organization"). (Glennan et al., 2022, P. 145)

Figure 4.1 offers a visual representation of how the mechanism, the mechanism's components, and the phenomenon are related. It is important to note that, as

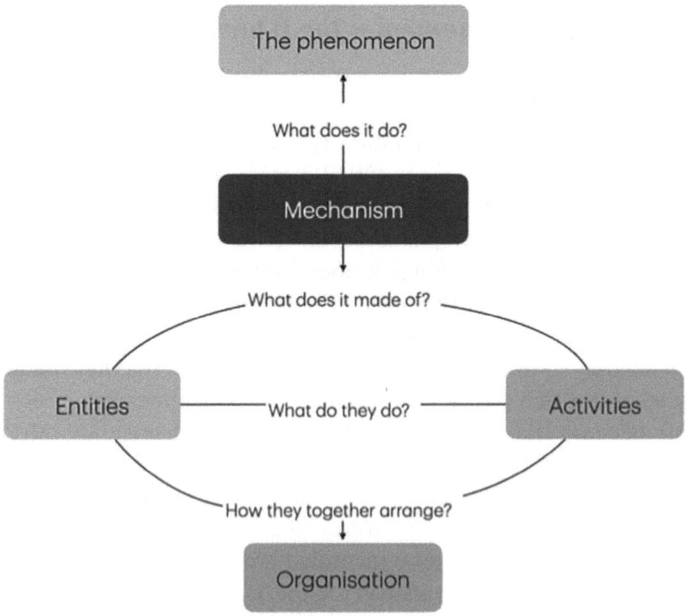

Fig. 4.1 Illari and Williamson's (2012) wide framework comprises three essential elements represented here: what the mechanism does (the "phenomenon"), what it is made of ("entities"), what its parts do ("activities"), and how they are structured ("organization")

described in the minimal definition, mechanisms "are bits of worldly stuff" (Glennan et al., p. 147, 2022) that make things happen, whereas models are the various ways we represent mechanisms, which can take different forms (mathematical, diagrammatic, etc.). While mechanisms exist in reality, we can only discover, describe, and explain them through models (Glennan et al., 2022). Explanations, particularly mechanistic ones, explain how phenomena happen by using mechanistic models to show the way mechanisms work (Bechtel & Abrahamsen, 2005; Glennan, 2017; Craver, 2007). Mechanistic models contrast with phenomenal models; whereas phenomenal models merely describe what happens, mechanistic models show how it happens by representing both the phenomenon and the mechanism responsible for it. Importantly, all models are inherently partial and idealised—as Glennan (2017) observes, "the only perfect model of a mechanism is the mechanism itself" (p. 66).

A virtue of having a wide framework for mechanisms is that it helps in identifying the essential elements of mechanism discovery across diverse fields, with widespread consensus that this unfolds in three steps:

> (a) to identify and describe the phenomenon to be explained, (b) to find the entities on the mechanism, and describe what they do (activities), and (c) to find out the organization of entities and activities by which they bring about the phenomenon (Illari & Williamson, 2012).

In the following, we will further describe each one of these three essential elements and how they are understood for wide applicability.

4.2.5 Identify the Phenomenon for Which the Mechanism Is Responsible For

A central claim in mechanistic literature is that mechanisms are defined by what they do (the phenomenon). Glennan (2017) argues that one cannot even identify a mechanism without specifying what it is that the mechanism does. "A mechanism is always a mechanism of a given phenomenon" (Craver & Darden, 2013). A mechanism is the mechanism of the things it does (Glennan, 2002). Hence, mechanisms and phenomena are reciprocally related. The mechanism of heartbeats is *responsible for* the heartbeat; the mechanism of melting is *responsible for* melting; the mechanism of light refraction is *responsible for* light refraction. Identifying and describing what it is that the mechanism does (the phenomenon) guides its discovery not only because it constrains the nature of the mechanism by informing what the mechanism should be able to do but also helps to establish its boundaries by determining what is included in the mechanism and what is not (Craver & Darden, 2013). Describing the phenomenon is a crucial step that shapes the space for discovery. The entities and activities to be identified are only relevant because, together, they do something that is responsible for the target phenomenon. Ultimately, a mechanistic explanation succeeds only if the mechanism it describes is responsible for the target phenomenon (Illari & Williamson, 2012). Section 3 will explore how identifying something as a phenomenon and describing it is not always straightforward. In the field of chemistry, for instance, this often involves conceptualising events that we observe in the world in terms of macroscopic constructs or models, such as pressure, temperature, volume, and so on (Taber, 2013; Talanquer, 2011). Additionally, some events span long periods, as seen in astronomy, which might make it challenging to establish them as phenomena (Bailer-Jones, 2009).

"Responsible for" is a broad term that encompasses the various ways in which a mechanism and a phenomenon can be related (Illari & Williamson, 2012). In the mechanistic literature, it is consensual that mechanisms can be *responsible for* their phenomenon in two ways: mechanisms cause or produce their phenomena and mechanisms constitute or underlie their phenomena (Craver & Darden, 2013; Glennan, 2017; Kaiser, 2018). In the first case, the phenomenon causally depends on the result of the outcomes of the mechanism's components. Whereas in the second case, the activity of the mechanism as a whole depends upon the activities of its components. Here, the phenomenon is something that is constituted rather than is caused by the mechanism's components. In Sect. 3, examples will be provided to illustrate these different kinds of phenomena.

4.2.6 Entities and Activities

The various proposals for understanding mechanisms gather consensus that all mechanisms consist of two kinds of components: the things that make them up, whatever they are called "entities" (Machamer et al., 2000), "parts" (Glennan, 2002), or "components parts" (Bechtel & Abrahamsen, 2005), and the things that those things do, whatever this would be activities, interactions, or operations.[2] Finding out the relevant entities and activities for the mechanism is an essential step in mechanism discovery that Bechtel and Richardson (2005) refer to as *decomposing the mechanism*. The idea of decomposition is consensual. However, it is important to carefully consider the term "decomposition" for its wide applicability, as it might intuitively suggest breaking something larger into smaller parts; not all mechanisms follow this type of decomposition (Glennan et al., 2022; Illari & Williamson, 2012). Some mechanisms are about the little explaining the big, although, in others, the big explains the small. We will provide some examples in the next section. In a wide framework, decomposing can be better understood as finding the relevant entities that make up the mechanism and describing what they do. Mechanisms can deal with a wide variety of entities, from extremely tiny ones, such as atoms, molecules, electrons, etc., to extremely large ones, such as stars, planets, and galaxies. Some mechanisms even lead to both, such as the end of a star, which is an interplay between the massive star's core and electrons' degeneracy (Illari, 2019). Also, as certain entities undergo minimal changes over time, others, like mRNA, are more

[2] Illari and Williamson favour using the terms "entities" and "activities" because "these terms provide rhetorical benefits for avoiding bias towards entities, and there is no restriction on the arity of activities" (2012, p. 126).

temporary. mRNA is produced from DNA, serves as a template for protein synthesis, and is broken down afterwards (Illari & Williamson, 2012). Finally, New Mechanists (e.g., Craver & Darden, 2013; Glennan, 2017; Kaiser, 2018; and Illari & Williamson, 2012) generally agree in not attributing any sort of hierarchy between activities and entities; both are integral parts of the mechanism and equally important for what the mechanism is responsible for—the phenomenon. An entity or group of entities has to do something, and an activity must be the activity of an entity or group of entities (Illari & Williamson, 2012). However, when explaining different kinds of phenomena, entities can sometimes be more interesting than the activities, and vice versa (Craver & Darden, 2013; Glennan & Illari, 2018). As we will see in Sect. 3, for some phenomena, such as photosynthesis and astronomical phenomena, each entity or group of entities is involved in specific activities. Whereas, for the thermal expansion of a gas, numerous similar entities are at play, and it is their collective actions that matter.

In a nutshell, mechanisms have two kinds of components: entities and activities. Entities can be of varying nature and sizes; some are parts of something bigger, others not; some entities are relatively stable, others more ephemeral; both entities and activities join the same status; they are parts of the mechanism.

4.2.7 Organization

We identify a target phenomenon, find its entities, and describe their activities, but that is not enough (Illari & Williamson, 2010). A mechanism is not merely the sum of its parts; the entities and activities, on their own, are not responsible for the phenomenon; rather, it is whatever they do (Craver, 2006). It is how the mechanism's components together organize that gives the final elements of how the phenomenon is brought about (Illari & Williamson, 2012). One can argue that the same activities and entities organized differently will produce something different. Glenna and Illari give us a good illustration: "Electrical circuits nicely illustrate this fact: the very same resistors or capacitors can exhibit very different resistance or capacitance depending upon whether they are wired in parallel or in series" (2018, p. 96).

How can we characterise *organization* to apply widely?— Organization, as Illari and Williamson refer to, is "whatever relations between the entities and activities discovered produce the phenomenon of interest: when activities and entities each do something, and do something together to produce the phenomenon" (2012, p. 128). At the most basic level, mechanisms exhibit structural forms of organization. This includes spatial organization, which encompasses the location, size, shape, and position of components, and temporal organization, which involves the order, rate, and duration of activities (Glennan, 2017; Kaiser, 2018). But, more broadly, organization is not only restricted to spatial or temporal forms; New Mechanists (e.g., Bechtel & Abrahamsen, 2005; Craver, 2006; Illari & Williamson, 2010) recognise the importance of other forms. Mechanisms can exhibit various forms of organization, such as feedback and control systems in biological mechanisms, where components regulate each other's behavior. For instance, homeostatic mechanisms maintain system stability by returning to their original state after perturbations. These forms of organization are significant because they illustrate how mechanisms can be organized through regulatory relationships and emergent properties rather than solely through spatial and temporal arrangements (Illari & Williamson, 2010). In its minimal definition, organization is permissive, allowing for various forms; for instance, in Sect. 3, we will show that random motion constitutes a form of organization in the mechanisms of thermal expansion of a gas.

4.3 A Wide Framework for Mechanism Discovery Across Scientific Domains: The Case of Seasons, Gases' Behavior, and Photosynthesis

In this section, we will apply Illari and Williamson's broad framework—as outlined in Sect. 2 (see also Fig. 4.1)—to characterise the mechanisms responsible for three distinct phenomena across various domains: the thermal expansion of a gas, the difference in daylight across seasons, and how plants grow. We have specifically chosen these phenomena because they are commonly found in science curricula worldwide across different grade levels (e.g., NGSS, 2013; NRC, 2012) and are suitable for mechanism-based explanations, which involve various types of entities, activities, and forms of organization. When describing each mechanism, we will provide an appropriate level of complexity for middle/high-grade levels.

4.3.1 The Case of the Thermal Expansion of a Gas

The phenomenon we will address is the expansion of a gas in a non-rigid container (let's say a balloon) when it is heated. Two main types of explanations can be offered: covering law and mechanistic. The covering law explanation utilises laws, such as the gas laws, to explain the behavior of the gas under certain conditions. These explanations have been widely accepted in physics as they generate quantitative predictions about the expected behavior of the gas (Glennan et al., 2022). They are also common practice in science classrooms, where students are not often pressed to pursue more sophisticated forms of explanations (Tang et al., 2020; see also Chap. 10.). However, covering law-type explanations does not show how and why the gas in the balloon behaves in such a way (Glennan et al., 2022; Woodward, 2003). For this, a description of the mechanism underlying the gas laws that predict the gas behavior is necessary. An account of this mechanism utilises the kinetic model of matter to explain how and why the volume of a gas increases with temperature.

4.3.1.1 Identifying and Describing the Target Phenomenon

We observe that a balloon expands when left outside in the Sun. However, to explain what causes this behavior and how it occurs, we need to attribute this observation to a specific phenomenon: the thermal expansion of a gas in a flexible container. A mechanistic explanation for the thermal expansion of a gas (or any other phenomenon) will only succeed if the mechanism described is the mechanism *responsible for* the identified features of the phenomenon (Illari & Williamson, 2012). Therefore, we must then provide a comprehensive description of the behavior of the target phenomenon using the intellectual tools of a particular field (Craver & Darden, 2013). In this case, it entails describing the behavior of the gas when heated in a flexible container in terms of conceptual constructs such as temperature, internal pressure, volume, and so on (Taber, 2013; Talanquer, 2011). Hence, a formal description of what causes the behavior of the target phenomenon would be that a balloon (a closed but flexible container) containing a certain amount of gas, when left outside in the sun, experiences an increase in its temperature, which leads to a rise in both internal pressure and the volume of the balloon.

4.3.1.2 Entities and Activities in the Case of Thermal Expansion of a Gas

In the case of the thermal expansion of a gas, one can give an account of how the phenomenon came to be (what causes the behavior of the target phenomenon); however, we also want to know about the mechanism underlying such behavior and how it works. Hence, we then need to identify its entities and describe their activities. In this case, the things at the level of the phenomenon/the higher level (the gas expanding as a whole) are constituted or made of smaller entities at the lower level (the mechanism's components). The gas in the container is made of tiny particles (atoms and/or molecules) that are not visible to the naked eye, a magnifying glass, or even a basic optical microscope. Identifying them, in this case, involves considering that something "bigger" (the gas in the container) is made of "smaller pieces" and then determining the activities of these smaller pieces. This has epistemic parallels to what Talanquer (see Chap. 10) refers to as "zoom in" from the macroscopic level to the multiple-particle scalar level. In contrast to the two examples we will present next, the gas is made of many similar entities; we are not interested in what they do individually, but rather as a whole; it is what they together do (randomly moving through the empty space) that matters. The entities, activities, and their organization relevant to the mechanism underlying the phenomenon of the thermal expansion of the gas in a balloon are described in Table 4.1.

4.3.1.3 Organization

Saying that the molecules are in random motion and evenly distributed through empty space describes a kind of organization (Glennan et al., 2022). Keeping a wide perspective, organization simply means how the entities and their activities

Table 4.1 The entities, activities, and organization relevant to the mechanism underlying the thermal expansion of a gas in a balloon

entities	*activities*
The *large number of molecules in the balloon* (expansible container)	The molecules *randomly move* through empty space The molecules are *evenly distributed* through empty space The molecules *collide* with each other and with the walls of the non-rigid container (they move) *Heating* the container The molecules *move faster* and the number of *collision increases*
How are entities and activities ***organized*** so that the mechanism is responsible for the phenomenon?	
The molecules are located inside the balloon and they move randomly throughout the empty space. Heating the balloon and the gas results in more kinetic energy, more powerful collisions, and increasing the pressure on the walls of the balloon, thus causing it to expand	

are arranged to bring about the phenomenon. The behavior of a gas at the phenomenon level emerged from this particular kind of organization: the continuous and random motion of the entities (the molecules in the balloon) that make up the things at the phenomenon level (Glennan et al., 2022). This sort of mechanism differs from others in that each different component part of the mechanism has a specific role in bringing about the phenomenon by virtue of its local relations to other components (Glennan et al., 2022). In the case of gas behavior, no specific entity has a specific role; rather, it's how they together randomly arranged (Fig. 4.2).

4.3.2 *The Case of the Seasons: the Difference in Daylight*

The event to be explained here pertains to seasonal phenomena, specifically the differences in daylight hours between the northern and southern hemispheres during winter in the northern hemisphere.

4.3.2.1 Identifying and Describing the Target Phenomenon

Establishing something as a phenomenon often involves identifying and describing a recurring pattern in the world (Bailer-Jones, 2009), implying repeated observation. Some phenomena observed only once cannot be established as such (Kraus, 2024). Seasonal phenomena are a striking case. From Earth's perspective, we experience longer days during the summertime. From March 21 to June 21, someone in the northern hemisphere will observe that each day, the Sun "rises" earlier and "sets" later, resulting in more hours of daylight. Simultaneously, during the same period, someone in the southern hemisphere will experience the opposite: shorter days (fewer hours of daylight).

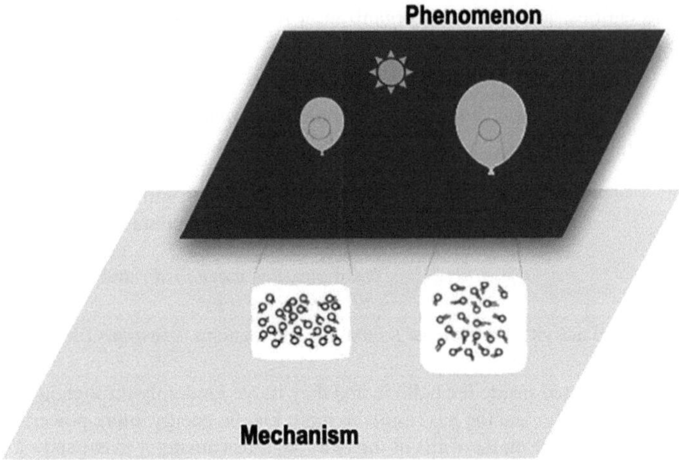

Fig. 4.2 A visual representation of the underlying mechanism for the thermal expansion of gas. At the top (macro-scale) is the behavior of the mechanism as a whole (the phenomenon): some system (the gas in the balloon) acts in some way (the balloon expands at the sun). At the bottom is the mechanism responsible for how this system behaves this way. "Zooming in" to the mechanism level (at the multi-particle scale), we found the entities and their activities and how they together organized. A mechanistic model for the thermal expansion of a gas shows both the horizontal (causal) and vertical (constitutive) dimensions, describing how the organized parts cause and underlie the target phenomenon

4.3.2.2 Entities and Activities in the Case of Seasonal Phenomena

The difference in daylight exemplifies a phenomenon that is a causal result of the activities and interactions of the mechanism's entities rather than being made of them. In this case, a few different entities engage in specific activities that together cause the phenomenon but do not constitute it. However, the phenomenon is still produced by the whole mechanism (Glennan, 2017).

Mechanistic explanations for astronomical phenomena do not involve zooming into the system to identify tiny entities that constitute and are responsible for the properties and behavior of the macroscopic elements we experience, as for gas behavior or many other phenomena in chemistry (e.g., see Chap. 10). However, it still involves looking inside a system and finding its relevant parts. The system to consider consists of large objects that are far apart and so not available to us from our Earth's perspective. Instead of zooming in for astronomical phenomena, such as seasons, tides, lunar phases, day-night cycles, etc., one has to zoom out from our Earth's perspective to a space's perspective (see Fig. 4.3). The epistemic work is, however, similar; it requires one to look inside the mechanism to search for relevant parts and their relevant activities that cause the target phenomenon (Bechtel & Abrahamsen, 2005). Table 4.2 describes the relevant entities, activities, and their organization responsible for the difference in daylight hours.

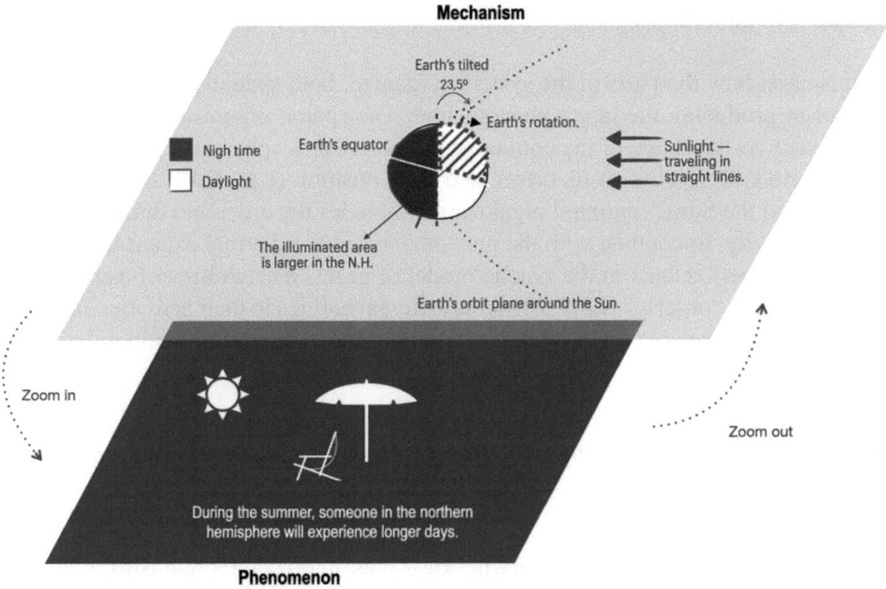

Fig. 4.3 A visual representation of the mechanism responsible for the differences in daylight across the seasons. From our Earth's perspective, we experience longer days during the summer. Identify the relevant entities, such as the Earth, the Earth's Tilt, the sun, and so on, and describe what they are doing (activities), such as rotating, travelling, orbiting, and involve zooming out to a space's perspective. Contrasting with the particle model for the expansion of a gas, here it is possible to represent each component of the mechanisms in detail

Table 4.2 The entities, activities, and their organization relevant to the phenomenon of the difference in the daylight hours in the northern and southern hemispheres

entities	*activities*
The *Sun*; The *Earth* and its *spherical* nature; The *Earth's tilt*; The Earth's illuminated area;	The Sun *emits* light (*sunlight*) *travelling* in straight lines. The Earth *rotates* on its axis every 24h; The Earth in its annual *orbit around* the Sun;
How are entities and activities ***organized*** so that the mechanism is responsible for the phenomenon?	
The Earth is tilted relative to its orbit around the Sun. During the summer, the northern hemisphere is tilted toward the Sun A larger portion of the Northern Hemisphere receives sunlight during this time of the year As the Earth completes its 24-hour rotation, locations in the northern hemisphere are exposed to more hours of daylight	

4.3.2.3 Organization

In this case, how the parts in the systems organize, both spatially and temporally, is critical in producing the target phenomenon. The spatial organization includes elements such as the shape of the entities (e.g., the Earth is spherical), their orientation (e.g., Earth's tilt relative to its orbit), and their position (e.g., Earth's position in its path around the Sun). Temporal organization includes the order and duration of each entity's activity. In contrast with the phenomenon of the thermal expansion of a gas, as well as others related to the kinetic model of gases, mechanisms of astronomical phenomena are "orderly" (Levi, 2014), as different entities in their activities each play a specific role in producing the phenomenon in virtue of their spatiotemporal relation.

4.3.3 The Case of Photosynthesis: How Do Plants Get Food and Grow?

Plants are primary producers, otherwise known as autotrophic organisms, that grow and survive without consuming other life forms. They can produce their own food through photosynthesis by harnessing energy (light) and nutrients available in their environment. This happens in the leaves of plants, inside the cells, in tiny organelles called chloroplasts, through chemical reactions.

4.3.3.1 Identifying and Describing the Target Phenomenon

Plant growth involves a process called photosynthesis. Hence, a mechanistic explanation for how plants grow requires describing the mechanism responsible for photosynthesis. The coupling of photosynthesis to plant growth is a critical step.

The mechanism of photosynthesis exhibits behaviors and changes. Plants grow and produce biomass, even in the absence of external food sources. This suggests an internal process capable of harnessing energy and producing organic matter. Also, the release of oxygen by plants, particularly during daylight hours, suggests the occurrence of a chemical reaction involving the absorption of light energy, an indicator of energy change.

4.3.3.2 Entities and Activities in the Case of Plant Growth

Plant growth mechanism has a hierarchical multilevel structure (see Fig. 4.4), characteristic of many mechanisms in the field of life sciences (Craver, 2007). Each level of the mechanism is made up of organized entities and activities. The behavior of the mechanism at a higher level is the product of its components at the lower levels. Therefore, decomposing proceeds by decomposing the behavior of a mechanism at a higher level into the behaviors of its component parts (entities and

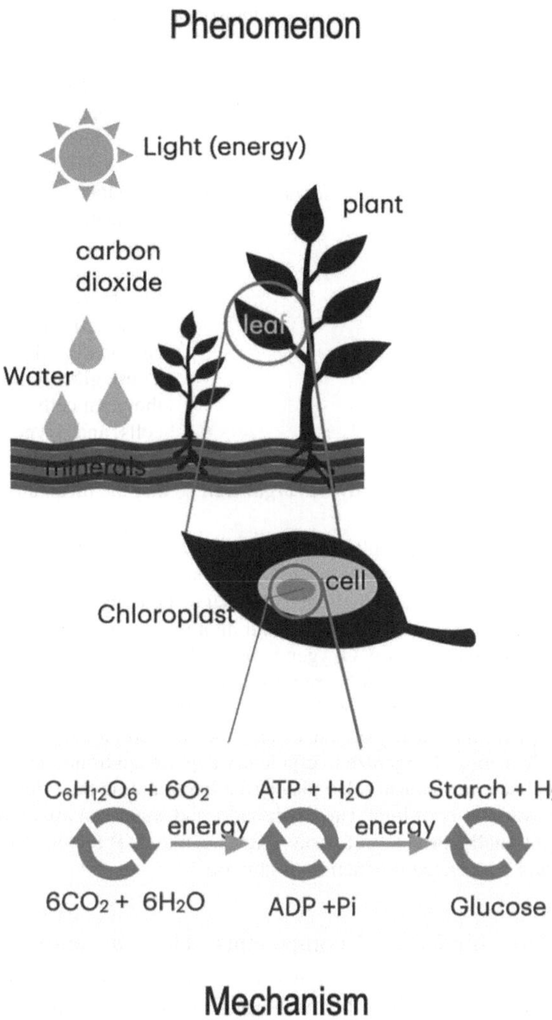

Fig. 4.4 A visual representation of the mechanism of plant growth. At the top of the mechanistic hierarchy, a plant is a system, and growing is an activity that the plant engages in. The plant has parts, e.g., leaves, stems, and roots. The parts engage in various activities, e.g., roots absorb water from the soil, leaves take in sunlight and absorb carbon dioxide from the air. These parts' activities are mechanistically constituted by their activities, e.g., the chemical reaction occurring in chloroplasts that use sunlight to transform water and carbon dioxide molecules into glucose and oxygen molecules

activities) at the lower level, decomposing the behaviors of the parts into the behaviors of their parts, and so on (Craver & Bechtel, 2007). Table 4.3 describes the relevant entities, activities, and organization responsible for the mechanism of plant growth.

4.3.3.3 Organization

Organising, in a mechanistic approach, establishes the ongoing conditions that enable the entities and activities to produce the target phenomenon (Craver, 2007). In the mechanism of photosynthesis, lower-level components are organized together

Table 4.3 The entities, activities, and their organization, relevant to the mechanism of plant growth

<table>
<tr><th>entities</th><th>activities</th></tr>
<tr><td>The leaves
The roots
The plant cells
The chloroplasts
The molecules of carbon dioxide, oxygen, water and glucose
The sunlight</td><td>Plants take in sunlight through their leaves
Chloroplasts in plant leaves capture sunlight;
Plants also absorb water from the soil (through the roots) and carbon dioxide from the air (through the stomata on leaves)
These nutrients enter specialised plant cells present in the leaves
Once inside the plant cells they are conducted to the chloroplasts, where photosynthesis occurs
Plants use sunlight (energy) to transform water and carbon dioxide into food (glucose) and oxygen
These glucose molecules and oxygen then leave the plant cells. Sugar is transported to all other cells, and oxygen is released into the atmosphere (through stomata on the leaves)</td></tr>
<tr><td colspan="2">How are entities and activities organized so that the mechanism is responsible for the phenomenon?</td></tr>
<tr><td colspan="2">Photosynthesis takes place in the chloroplasts of plant cells. Here, pigments absorb light (energy), which is used to separate the water molecules into their constituent atoms (chemical reaction), oxygen and hydrogen, and harvest energy in the form of a molecule designated by ATP. Hydrogen and ATP then power another biochemical cycle in which carbon dioxide is converted into glucose. Oxygen is then released into the atmosphere. Complex forms of organization, such as self-organization and feedback, are vital for the production of the phenomenon of plant growth. The mechanism of the process of photosynthesis involves a complex set of highly organized and coordinated biochemical reactions. Components of the mechanism self-organize to efficiently capture environmental nutrients and use energy. Feedback mechanisms also play a role in regulating the efficiency of the process. For instance, the availability of light, carbon dioxide, and water can affect the rate of photosynthesis. Additionally, products of photosynthesis, like ATP, feedback into the process to regulate and ensure a balanced production of glucose</td></tr>
</table>

to form higher-level components. The way entities and activities are organized determines how lower-level components become higher-level components. Organization is the interlevel relation between a mechanism as a whole and its components (Craver, 2007).

The mechanism of photosynthesis is an important example of a multi-level mechanism that consists of "mechanisms within mechanisms" (Franovic et al., 2023). Also, it involves entities and activities at multiple levels and non-trivial forms of organization. Additionally, it requires the use of chemistry science ideas about invisible changes in matter and energy to explain life science phenomena (Roseman et al., 2021). The description of a mechanism can lie anywhere on a continuum between a *mechanism sketch* and an *ideally complete description of the mechanism* (Craver & Darden, 2013). A mechanism *sketch* is an incomplete description of a mechanism. It identifies some entities, activities, and features of the mechanism's organization, but it has gaps Craver & Darden, 2013). These gaps can be made evident in a *mechanism sketch* by black boxes. They sign the presence of a "unit" within the mechanism, which the entities and activities are yet to be unpacked

(Haskel-Ittah, 2023, see also Chap. 10). Teaching a complete mechanism of photosynthesis would be unreasonable, unproductive, and time-consuming at the middle school level (Haskel-Ittah, 2023). Instead, Haskel-Ittah suggests that the mechanism of photosynthesis should first be introduced as a black box, with the input of light and carbon dioxide and the output of oxygen and glucose. As the author stressed, this black box should be recognised by students as something, there, yet to be unpacked. As students progress to higher levels, some components of the mechanism may be unpacked, while others remain as black boxes to be understood later.

The three cases mentioned above do not fully encompass the vast array of mechanisms found in various scientific fields, but they do demonstrate that mechanistic explanations in many disciplines can be unified under a wide framework. Essentially, all mechanisms consist of two types of components: *entities and activities* that we need to find. Entities and activities can sometimes be the constituent parts of a mechanism underlying a phenomenon. For example, a gas expanding in a container is made of a large number of similar entities moving randomly. In other cases, entities and activities that constitute a mechanism at a higher level can be, themselves, a mechanism at a lower level, such as cells in a plant's leaves. Whereas other phenomena, such as the Seasons, are caused by entities and their activities organized in a whole system but not made of them. Entities and activities hold equal importance (Illari & Williamson, 2012). However, for various phenomena, sometimes entities are more relevant than activities or the other way around. For example, in photosynthesis, the detailed structure of entities is critical. Whereas, in mechanisms explaining ideal gas behavior, entities are assumed to be similar, making activities and organizations vital. In addition, *organization* is an essential element of any mechanism. Entities and activities on their own do not bring about phenomena; rather, it is how they, together, organize. Within a wide framework, *organization* is permissive, encompassing various forms that are not limited to spatiotemporal relations.

Although we have not included examples from other fields, such as social sciences, in the current chapter, we want to emphasise that mechanism-based explanations are ubiquitous across many disciplines (Glennan et al., 2022), and they share numerous commonalities that we are describing here. Furthermore, many mechanistic explanations are built using components from different fields (Illari & Williamson, 2012).

4.4 Final Remarks Comments

In this chapter, based on current work on the philosophy of science (Illari et al., 2012, 2022), we present and discuss a definition of mechanisms that apply widely across science domains. This wide definition offers what is common to mechanisms across various domains. Our main purpose was to highlight the ubiquity of mechanistic explanations across different fields, from life to physical sciences, and what they share in common. By recognising mechanistic reasoning as a cross-cutting concept (NGSS, 2013), we are taking it as a way of thinking that widely applies across science domains but also, desirably, in everyday life, as Ryan et al. are showing in Chap. 12. Therefore, we need to fundamentally understand what this way of thinking shares in various fields so that we can develop widely applicable approaches and tools that enable students to learn and adeptly apply mechanistic reasoning as a valuable intellectual resource to make sense of, understand, and explain complex real-world phenomena.

Characterising mechanisms within a broad framework does not overlook the importance of considering the many specificities encountered in different domains, as discussed in Chap. 10 regarding chemical mechanisms, inter alia; see Chaps. 1 and 3. Instead, the aim is to look for what we have in common across different domains when thinking about mechanisms and building mechanistic explanations. This has the potential to offer us an overarching rationale useful for grasping, making sense of, analysing, and characterising the specificities that are at work across these various fields. However, it is important to clarify that we do not regard this framework as an approach for teaching mechanistic reasoning or as scaffolding for students to build mechanistic accounts; rather, we view it as a way of looking for mechanistic reasoning in science education, contributing to the overlapping of our ongoing work, understandings, and discussions. We contend that a wide understanding of mechanisms, as explored in this chapter, can serve as a foundation for guiding future research that envisions mechanistic reasoning as a valuable and highly transportable way of thinking that students can use across various domains in school and in everyday life.

References

Andrade, V., Freire, S., & Baptista, M. (2019). Constructing scientific explanations: A system of analysis for students' explanations. *Research in Science Education, 49*, 787–807. https://doi.org/10.1007/s11165-017-9648-9

Andrade, V., Shwartz, Y., Freire, S., & Baptista, M. (2022). Students' mechanistic reasoning in practice: Enabling functions of drawing, gestures and talk. *Science Education, 106*, 199–225. https://doi.org/10.1002/sce.21685

Bailer-Jones, D. M. (2009). *Scientific models in philosophy of science*. University of Pittsburgh Press.

Bechtel, W. (2009). Looking down, around, and up: Mechanistic explanation in psychology. *Philosophical Psychology, 22*(5), 543–564. https://doi.org/10.1080/09515080903238948

Bechtel, W., & Abrahamsen, A. (2005). Explanation: A mechanistic alternative. *Studies in History and Philosophy of the Biological and Biomedical Sciences, 36*, 421–441. https://doi.org/10.1016/j.shpsc.2005.03.010

Craver, C. (2006). When mechanistic models explain. *Synthese, 153*, 355–376. https://doi.org/10.1007/s11229-006-9097-x

Craver, C. (2007). *Explaining the brain*. Clarendon.

Craver, C., & Bechtel, W. (2007). Top-down causation without top-down causes. *Biology and Philosophy, 22*, 547–563. https://doi.org/10.1007/s10539-006-9028-8

Craver, C., & Darden, L. (2013). *In search of mechanisms: Discoveries across the life sciences*. University of Chicago Press.

Craver, C., Tabery, J., & Illari, P. (Fall 2024 Edition) Mechanisms in science. In Edward N. Zalta & Uri Nodelman (Eds.), *The Stanford encyclopedia of philosophy*. https://plato.stanford.edu/archives/fall2024/entries/science-mechanisms

Franovic, C. G.-C., Williams, N. R., Noyes, K., Klymkowsky, M. W., & Cooper, M. M. (2023). How do instructors explain the mechanism by which ATP drives unfavorable processes? *CBE—Life Sciences Education, 22*(50), 1–16. https://doi.org/10.1187/cbe.23-05-0071

Glennan, S. (2002). Rethinking mechanistic explanation. *Philosophy of Science, 69*(3), S342–S353.

Glennan, S. (2017). *The new mechanism philosophy*. Oxford University Press.

Glennan, S., & Illari, P. (2018). Varieties of mechanisms. In S. Glennan & P. Illari (Eds.), *The Routledge handbook of mechanisms and mechanical philosophy* (pp. 91–103). Routledge.

Glennan, S., Illari, P., & Weber, E. (2022). Six theses on mechanisms and mechanistic science. *Journal of General Philosophy of Science, 53*, 143–161. https://doi.org/10.1007/s10838-021-09587x

Haskel-Ittah, M. (2023). Explanatory black boxes and mechanistic reasoning. *Journal of Research in Science Teaching, 60*(4), 915–933. https://doi.org/10.1002/tea.21817

Illari, P. (2019). Mechanisms, models and Laws in understanding supernovae. *Journal for General Philosophy of Science, 50*, 63–84. https://doi.org/10.1007/s10838-018-9435-y

Illari, P. M., & Williamson, J. (2010). Function and organisation: Comparing the mechanisms of protein synthesis and natural selection. *Studies in History and Philosophy of Biological and Biomedical Sciences, 41*(3), 279–291. https://doi.org/10.1016/j.shpsc.2010.07.001

Illari, P. M., & Williamson, J. (2012). What is a mechanism? Thinking about mechanisms across the sciences. *European Journal for Philosophy of Science, 2*, 119–135. https://doi.org/10.1007/s13194-011-0038-2

Kaiser, M. I. (2018). The components and boundaries of mechanisms. In S. Glennan & P. Illari (Eds.), *The Routledge handbook of mechanisms and mechanical philosophy* (pp. 116–128). Routledge.

Kraus, S. F. (2024). The method of observation in science education: Characteristic dimensions from an educational perspective. *Science & Education, 33*, 1033–1068. https://doi.org/10.1007/s11191-023-00422-x

Krist, C., Schwarz, C. V., & Reiser, J. B. (2019). Identifying essential epistemic heuristics for guiding mechanistic reasoning in science learning. *Journal of the Learning Sciences, 28*(2), 160–205. https://doi.org/10.1080/10508406.2018.1510404

Levi, A. (2014). Machine-likeness and explanation by decomposition. *Philosophers' Imprint, 14*(6), 1–15.

Machamer, P. (2011). Phenomena, data and theories: A special issue of synthese. *Synthese, 182*, 1–5. https://doi.org/10.1007/s11229-009-9622-9

Machamer, P., Darden, L., & Craver, C. F. (2000). Thinking about mechanisms. *Philosophy of Science, 67*(1), 1–25.

NGSS Lead States. (2013). *Next generations science standards: For states, by states*. National Academies Press.

NRC (National Research Council). (2012). *A framework for K-12 science education: Practices, crosscutting concepts, and core ideas*. National Academies Press.

OECD. (2016). *Definition and selection of competencies (DeSeCo)*. Retrieved from www.oecd.org/edu/skills-beyond-school/definitionandselectionofcompetenciesdeseco.htm

Roseman, J. E., Koppal, M., Abell, C. H., Pappalardo, S., & Schiff, E. (2021). Photosynthesis: Matter and energy for plant growth. In J. Nordine & O. Lee (Eds.), *Crosscutting concepts: Strengthening science and engineering learning* (pp. 269–294). NSTA Press.

Russ, R. S., Scherr, R. E., Hammer, D., & Mikeska, J. (2008). Recognising mechanistic reasoning in student scientific inquiry: A framework for discourse analysis developed from philosophy of science. *Science Education, 92*(3), 499–525. https://doi.org/10.1002/sce.20264

Ryan, Z., Danish, J., Zhou, J., Stiso, C., Murphy, D., Duncan, R., Chinn, C., & Hmelo-Silver, C. E. (2023). Investigating students' development of mechanistic reasoning in modeling complex aquatic ecosystems. *Frontiers in Education, 8*, 1159558.

Taber, K. S. (2013). Revisiting the chemistry triplet: Drawing upon the nature of chemical knowledge and the psychology of learning to inform chemistry education. *Chemistry Education Research and Practice, 14*(2), 156–168. https://doi.org/10.1039/C3RP00012E

Talanquer, V. (2011). Macro, submicro, and symbolic: The many faces of the chemistry "triplet". *International Journal of Science Education, 33*(2), 179–195.

Talanquer, V. (2018). Exploring mechanistic reasoning in chemistry. In J. Yeo, T. W. Teo, & K. S. Tang (Eds.), *Science education research and practice in Asia-Pacific and beyond* (pp. 39–52). Springer.

Talanquer, V. (n.d.). *The nature of mechanistic reasoning in chemistry*. (in this book).

Tang, X., Elby, A., & Hammer, D. (2020). The tension between pattern-seeking and mechanistic reasoning in explanation construction: A case from Chinese elementary science classroom. *Science Education, 104*, 1071–1099. https://doi.org/10.1002/sce.21594

van Mil, M. H. W., Boerwinkel, D. J., & Waarlo, A. J. (2013). Modelling molecular mechanisms: A framework of scientific reasoning to construct molecular-level explanations for cellular behaviour. *Science & Education, 22*(1), 93–118. https://doi.org/10.1007/s11191-011-9379-7

Woodward, J. (2003). *Making things happen: A theory of causal explanation*. Oxford University Press.

Part II
Promoting Mechanistic Reasoning—Implications for Supporting Mechanistic Reasoning

Chapter 5
Facilitating Mechanistic Reasoning with Gesture-Augmented Simulations: A Study of Two Students' Sensemaking of Air Pressure

Nitasha Mathayas

Mechanistic reasoning (MR), the process of identifying underlying entities and activities that cause observable scientific phenomena (Krist et al., 2019; Russ et al., 2008), is a hallmark of scientific discipline. This type of thinking gained prominence during the era of Galileo and Newton, laying the groundwork for modern science (Machamer et al., 2000). Remarkably, psychologists and learning scientists find that even young children naturally demonstrate this form of reasoning (Bolger et al., 2012; Erb et al., 2013; Schlottmann, 1999). However, a distinguishing feature of scientific MR is that often, the entities involved are *unobservable* and often *theoretical* entities (Brown, 2018; Clement, 2013). As a result, students must make significant conceptual leaps from what they observe to abstract, underlying mechanisms.

Since MR exemplifies spatial reasoning and mental animation, leveraging bodily action to make sense of the movements and interactions of abstract entities can support learner's MR (DeSutter & Stieff, 2017). Specifically gesturing with the hands can play a valuable role in facilitating learners' MR. As part of the processes of constructing mechanistic explanations, in this chapter I will describe how representational gesturing with the hands fosters the learners' mechanistic reasoning. By

This material is based upon work supported by the National Science Foundation under Grant No. DUE-1432424.NSF funded grant work. Project GRASP is a collaboration between the University of Illinois at Urbana-Champaign and the Concord Consortium. The author would like to thank the principal investigators Dr. Robb Lindgren, Dr. David E. Brown, and Nathan Kimball, and fellow research associate Dr. Robert Wallon for their role in shaping this work and advising the research conducted.

N. Mathayas (✉)
University of Illinois at Urbana-Champaign, Champaign, IL, USA
e-mail: mathaya2@illinois.edu

N. Graulich et al. (eds.), *Exploring Mechanistic Reasoning in Science Education*, Contributions from Science Education Research 17,
https://doi.org/10.1007/978-3-032-10246-1_5

describing a pair of middle school students making sense of air pressure in a sealed syringe with the help of a gesture controlled computer simulation, I will show how directed gesturing in a visually rich, collaborative augmented reality environment facilitates students' MR about air pressure. Drawing on insights from this case study, I will conclude with implications for leveraging gestures to support MR in science classrooms.

5.1 A Practical Framework for MR

The past two decades have seen an increased focus on mechanisms in science with philosophers debating both explanatory roles and the underlying organizational structures of mechanisms (Glennan et al., 2022). Although many characteristics of mechanisms are yet to be resolved, they are understood to be a foundational form of scientific thinking aiding in scientific discovery and causal explanation. A minimal definition of mechanisms is that, "a mechanism for a phenomenon consists of entities (or parts) whose activities and interactions are organized so as to be responsible for the phenomenon" (Glennan et al., 2022, p. 145). De Andrade, Schwartz, and Rodrigues (Chap. 4, this book) provide a framework applicable to a wide range of scientific topics typically taught in schools. Their framework foregrounds three features: (a) a *description* of the phenomenon (b) *identifying* the entities and their related activities, and (c) the *organization* of these entities and activities that are responsible for the phenomenon. In this study, I drew on the mechanistic reasoning (MR) heuristics defined by Krist et al. (2019), which align with the minimal definition and are reflected in the recent definitions discussed above. Their framework examines how students learn to recognize, choose, and engage in MR across scientific disciplines. The authors identified three key epistemic heuristics. First, thinking across scalar levels is crucial, as the entities responsible for a phenomenon typically exist at least one scalar level below the observable phenomenon. Second, students must make sense of the entities themselves—their regularities, properties, rules, and activities that may cause the phenomenon. Third, students need to coordinate relationships and develop a causal narrative that connects the underlying scalar level to the observable phenomenon. Using this framework in my case analysis, I will demonstrate how the collaborative, gesture-augmented environment facilitated these epistemic heuristics. First, however, I will explain how gesturing supports MR.

5.2 The Embodied Underpinnings of Mechanistic Reasoning

Bachtiar et al. (2022) recently reviewed the literature on MR and found that students often struggled with identifying the underlying entities and their associated activities across scalar levels. For example, middle school students may identify that the causes of thermal conduction occur at the molecular level but then associate friction between molecules as the source of heat transfer rather than molecular collisions (Mathayas et al., 2021). I argue that these challenges occur because students aren't accessing the appropriate source analogs to draw their thinking from. They may also struggle to map the appropriate relational structures from source to target domain. Using visual representations and computer simulations can help students visualize unobservable entities (Honey & Hilton, 2011). However, students often need support linking appropriate activities across observable and unobservable scales.

Rooted in constructivist epistemology, analogical reasoning is a core mechanism for constructing conceptual knowledge (Amin, 2015). An analogy is a mapping of knowledge from one domain to another, which conveys that a system of relations that holds among the source domain also holds in the target domain (Gentner, 1989; Hofstadter, 2001). Many argue that the human conceptual system is a series of analogical mappings from one domain to the other that is ultimately grounded in our physical experiences in the real world (Barsalou, 2008; Johnson, 2009; Lakoff, 1993). This implies that good instructional analogies are built upon embodied sources that leverage foundational image schemas of space, place, time and movement (Niebert et al., 2012).

Analogical and mechanistic reasoning are linked by the relational and structural mappings across domains. This implies that when MR involves unobservable entities and activities, there are underlying embodied analogs that can be leveraged to support learner's reasoning (Niebert & Gropengiesser, 2015). Moreover, MR specifically involves visual and spatial reasoning, that lends itself well to kinesthetic forms of science learning (DeSutter & Stieff, 2017).

In this article, I use the term "explanatory model" (Cheng & Brown, 2010) to address the kind of MR that involves unobservable entities. Explanatory models are imagistic mental models that depict how actions of unobservable entities involved in a phenomenon function together to make the observable phenomenon occur. For instance, an explanatory model of air pressure involves a molecular model of air that involves dynamic movement and elastic collisions of air molecules. However, as this example shows, an explanatory model of air molecules involves imagery and spatial reasoning to mentally simulate the unobservable events occurring in the phenomenon (Clement, 2009).

Mental models are complex structures. In our work, we have found that a productive way of thinking about a student's explanatory model is to unpack it along the lines of conscious and intuitive parts (Brown, 2018; Mathayas et al., 2021). The conscious part of the learner's model consists of verbal-symbolic knowledge and imagistic knowledge. These conscious elements of the learner's explanatory model can be externalized through a range of semiotic resources—such as drawings, gesturing, and speech—that the learner has at their disposal. However, the explanatory model also consist of intuitive or unconscious elements that a learner may not express but that guide their MR. These intuitions stem from our sensorimotor experiences and create mental frameworks called schematic structures. For instance, the containment schema—which encompasses concepts of boundaries, inside, and outside—can be derived from various experiences (see Niebert & Gropengießer [2015] for an analysis of climate change conceptions using this schema). As I will show next, gesturing with the hands has a distinct capability of representing both conscious and intuitive aspects of the learner's explanatory model. Gesturing is a way to dynamically represent an imagistic conception, but often when used spontaneously while speaking, the learner's gestures reveal intuitive conceptions that often precede conscious understanding from the learner (Cook et al., 2017). In other words, leveraging the modality of gesture in instructional settings has the potential to promote both deep and surface level conceptual change.

5.3 Gesturing in Science Learning

Hand gestures that accompany speech have many benefits in thinking, speaking, and problem solving such as reducing working load memory (Cook et al., 2012, 2017; Goldin-Meadow, 2011; Goldin-Meadow et al., 2001), as a semiotic resource grounding intellectual activity with the environment (Goodwin, 2000; Roth, 2001; Roth & Lawless, 2002), and representing imagery in thought (Hostetter & Boncoddo, 2017; McNeill, 1992). Gesture researchers identify categories of hand movements that have cognitive value. Unlike typical hand movements, these gestures have a defined beginning and end, consisting of phases such as rest, preparation, stroke, and return to rest (Roth, 2001). McNeill (1992) categorized these gestures into four kinds: (a) beat gestures: temporal moves with emphatic structure, (b) deictic gestures that are used in concrete and abstract pointing, (c) iconic gestures and (d) metaphorical gestures that are hand movements that bear a perceptual and semantic relation with concrete entities and events. For instance, using a fist to represent a ball would constitute an iconic gesture while a fist as a molecule would be a metaphorical gesture. Since both iconic and metaphorical gestures involve semantic and relational mapping, researchers often combine them and refer to them as *representational gestures* (Roth, 2001).

Two prominent accounts explain how we learn with gestures: a *functional* account (Novack & Goldin-Meadow, 2017) and a *mechanistic* account (Kita et al., 2017). The *functional* account explores *why* gesturing serves the functions that it does, and it underscores how representational gesturing is different from simple actions. This account helps with distinguishing the functions of gesturing in learning and communicating (Goldin-Meadow et al., 2009; Novack et al., 2014, 2018), and accounts for individual differences in gesturing (Özer & Göksun, 2020). The main claim is that representational gesturing is both an action and an abstraction away from action. By abstracting actions, gesturing aids in the generalization of concepts and enhances retention over time (Novack & Goldin-Meadow, 2017).

On the other hand, the *mechanistic* account explores *how* gesturing emerges from the human perceptual-motor system, which helps in understanding the root sources of abstract thought. For instance, language, composed of linguistic objects like words, clauses, and sentences, has its roots in sensorimotor experiences (Lakoff,

1993). Similarly, mental imagery retains the spatial, physical, and motoric representations of the event and directly takes information from sensorimotor simulations (Hegarty, 2004). Hostetter & Alibali (2008) postulate that gestures "emerge from the perceptual and motor simulations that underlie embodied language and mental imagery" (p. 502). Kita et al. (2017) hypothesize that gestures schematize information that then helps conceptual development. Through schematization, gesturing helps speakers activate, manipulate, package, and explore spatial-motoric information for the purposes of speaking and thinking.

Together, these accounts about gesturing have important educational implications. Exploring the role of gestures in educational settings, Singer (2017) synthesized the field to highlight two ways that gestures supported classroom learning: (1) gesturing helped learners visualize abstract concepts and, (2) gesturing is used in the co-construction of shared representations.

In science classrooms, teachers and students gesture with their hands all the time but studies show that representational gesturing facilitates students' science talk by representing the phenomenon non-verbally when students don't have the words to describe their observations and by supporting students' as they develop scientific discourse (Crowder, 1996; Givry & Roth, 2006). The ways students physically orient themselves around their gestures makes evident the ways they frame scientific explanations as for communication or personal sensemaking (Crowder, 1996; Kapon, 2017). The instructor's gesturing can also impact classroom learning as instructors use gestures to integrate and control the timing of different material resources they use in the classroom, such as linking drawings on a board with three-dimensional models to communicate concepts to their students (Flood & Harrer, 2023; Nathan et al., 2017; Pozzer-Ardenghi & Roth, 2007). Indeed, Givry and Roth (2006) argue for viewing conceptual development in classrooms as dialectical units involving all relevant semiotic resources that learners use during their meaning making processes. Gestures can be used as a form of assessment of student thinking particularly young children (Lira & Stieff, 2018; Tu et al., 2021). And the role of gestures for emerging English learners cannot be understated. Multilingual learners "bridge" conceptual thought with gestures to talk about science content (Belhiah, 2013; Ünsal et al., 2018).

While the previous studies have investigated student gestures in everyday settings, some studies investigated the impact of *directing* students to gesture while learning (Lindgren, 2015). Directed gestures are a set of conceptually relevant representational gestures that instructors direct students to perform during sensemaking (Padalkar & Ramadas, 2011; Plummer, 2009). Studies have found that encouraging students to use directed gestures activated motor systems that enhanced conceptual learning (Lindgren et al., 2016; Nathan & Martinez, 2015). Also, encouraging multilingual learners to gesture in their non-dominant language was found to support meaning construction (Ma & Jin, 2022).

Research on how gestures support mechanistic reasoning in science education is limited. However, the existing studies consistently demonstrate that gesturing helps students mentally visualize and animate mechanistic components. Research by Bolger et al. (2012) demonstrated that young children employed gestures to mentally animate the mechanical linkages of the simple machines. Postsecondary students enhanced their understanding of geological concepts by using gestures to represent layered rock formations (Atit et al., 2015), while physics students at the high school level gestured during moments they were mentally animating concepts (Stephens & Clement, 2010). de Andrade et al. (2022) investigated middle school students' MR as they constructed a model for a chemical phenomenon and concluded that "animating a model with gestures" may be an additional epistemic heuristic for MR.

In prior work, my colleagues and I showed that directed gesturing not only helped students animate the unobservable aspects of the mechanism. Rather, the gestures fostered new understanding by helping students activate relevant embodied schemas within the context of the observable phenomenon (Mathayas et al., 2021). Through this work, we argued that directed gesturing is not only a representational medium working within conscious aspects of students' explanatory models. We demonstrated that gesturing mediated the unconscious dimensions of students' explanatory models, making visible to researchers the embodied schemas of collision or friction that underpinned students' conscious explanations.

In what follows, I broaden these insights by showing how directed gesturing supports a pair of students as they collectively reason about a phenomenon. The dyadic interactions presented in this case provide a more realistic representation of how MR with directed gesturing might occur in classroom settings. Additionally, this case expands our understanding of directed gesturing—demonstrating how it not only provides cognitive support for the pair but also becomes integrated with the other material resources students employ during MR.

5.4 Case Study: Reasoning About Air Pressure Using a Gesture-Augmented Simulation

To illustrate how directed hand gestures fostered students' MR, I present a case study of a pair of middle school students who were asked to explain how air pressure builds up within a sealed syringe when the plunger is pushed inwards. These students were video recorded as part of a technology design research study conducted between 2014 to 2019 that involved developing and testing web-based science simulations augmented with hand gesture control [https://emit.education.illinois.edu/projects/grasp]. These students interacted with a gesture-augmented computer simulation to make sense of the essential mechanisms of air pressure. In this interview, I was the researcher working with Nina and Adrianna (both pseudonyms) who were in middle school in 2018. During the 40-minute interview, I first presented the pair with the target question about the syringe. I then guided them in using directed gestures to control the simulation (Fig. 5.1) and provided them 20 minutes of independent work time. During this period, they controlled the simulation together while following a sequence of tasks outlined on a worksheet. The session concluded with their final explanations and a debrief about their experience.

Throughout the interview, they were asked to verbally explain the phenomenon at several points.

Figure 5.1 is a screenshot of the simulation depicting air molecules moving randomly in an enclosed space. The syringe's plunger appears as a dark gray wall on the left side of the simulation. Users can interact with this wall in two ways: by clicking arrow buttons on the screen to move it inward or outward, or by using directed gestures. For this study, students controlled the plunger by making a specific gesture—tapping their fist against their palm. The fist represents a molecule/s in the simulation and the open palm represents the movable wall. Thus, by bumping their fist into their palm, the students represent the air molecules hitting the movable wall and depicting the pressure against the wall.

As an embodied analogical model, the hands and the associated actions of bumping into the wall were the source domain. Meanwhile, the target domain was the molecules depicted in the simulation. Our learning hypothesis was that students would be able draw on their embodied intuitions of movement and bumping while using these control gestures and those congruent action schemas would map to the abstract domain of molecular entities thereby facilitating their MR.

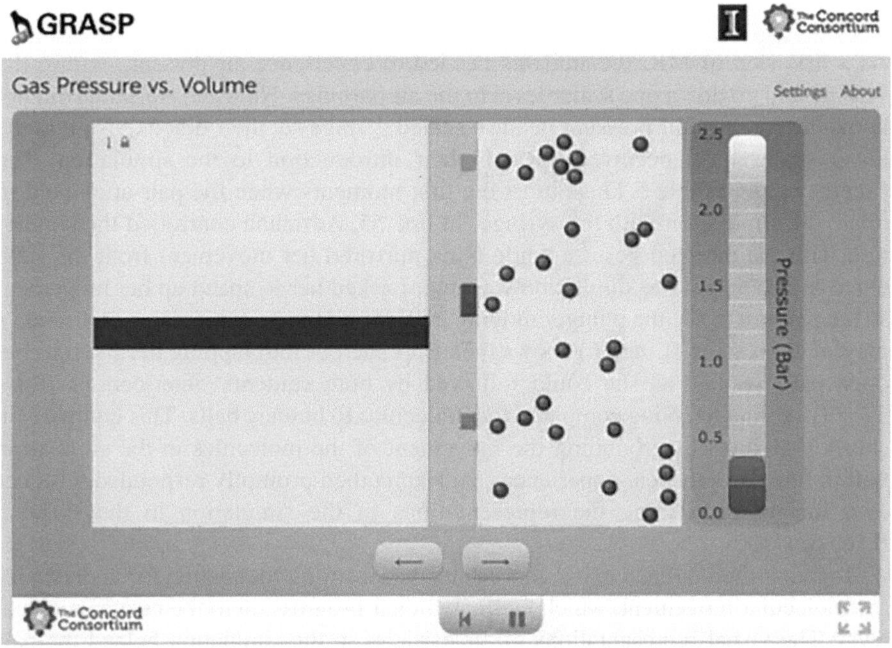

Fig. 5.1 The gas pressure versus volume simulation shows air molecules moving within an enclosed space. As the plunger moves inward, the rate of molecules hitting the plunger increases, indicated by red dots on the wall. The pressure bar on the right rises accordingly. Users can control this simulation either by clicking the arrows (shown above) with a mouse or trackpad, or—if they have a Leapmotion device—through hand gestures, by rapidly tapping one's fist against their palm to represent molecules colliding with the wall

This case was selected for this study because the two students demonstrated strong collaboration, thinking aloud together while building upon each other's ideas. This makes their case highly illustrative of their MR. Using methods of interaction analysis (Jordan & Henderson, 1995), I analyzed the video and transcript to explore the pair's interactions, identifying moments when their MR emerged in alignment with Krist et al.'s (2019) three heuristics: H1—thinking across scalar levels, H2—identifying and unpacking relevant factors, and H3—linking to coordinate relationships over time and space. When analyzing for the heuristic of identifying and unpacking relevant factors, the process of unpacking lower-level factors involved several distinct steps for students to complete. Therefore, I unpacked this level into two main activities: (a) identifying entities and their activities, and (b) identifying and reifying the causal mechanism. Next, across four phases of their interview, I will describe key interactions and how directed gesturing facilitated their MR.

5.5 Enactment of H1: Going down a Scalar Level and Activating Relevant Metaphors

As a first step of MR, the students needed to experience air pressure within the syringe and go down one scalar level to the air particles. Nina and Adrianna did not know the cause of air pressure inside a sealed syringe so, their first exposure to the lower scalar level occurred through their introduction to the simulation. The excerpt below (Table 5.1) captures the first moments when the pair attempted to relate the simulation with the syringe. In line 55, Adrianna controlled the simulation with the directed gesture while Nina mirrored her movement from the side. Once Adrianna got the simulation working, I asked her to speed up her movement so the pair got to see the plunger moving inward and the pressure rising. Adrianna's playful expression (Line 56) shows the effort she put into tapping her fist into her open palm as fast as she could followed by both students' interjections (lines 57–59). At line 61, Nina compared the molecules to bouncy balls. This comparison shows that she was grounding the movement of the molecules in the simulation within relevant physical experiences. Adrianna then promptly responded with her own insight connecting the representations in the simulation to the syringe. (Line 62).

Together their collective insights established a strong foundation for understanding molecular movement, which facilitated their progression to the next conceptual level. The visual representations of air particles in the simulation helped the pair move across scalar levels to begin the sensemaking work required of unpacking entities and interactions. This sensemaking process began immediately when Nina compared the air molecules to bouncy balls, providing an initial analogy for understanding. Adrianna then used pointing gestures to connect the simulation to the syringe, which helped ground the simulation for Nina and establish shared understanding between them.

Table 5.1 Nina and Adrianna figuring out how to control the simulation

Line	Speaker	Transcript	Video capture
55	Adrianna:	[Reading instructions on the screen] You have to turn it and then you have to push it	
56	Researcher:	Go faster [Adrianna responds with an exaggerated expression as she speeds up]	
57	Nina:	That's lit	
58	Researcher:	Slow down...[Adrianna slows her gesture]	
59	Adrianna:	I feel like I'm like— whoa! Oh, that's so cool	
61	Nina:	Those look like bouncy balls	
62	Adrianna:	Ah! that's the pressure in the ***thing***[a] [points to the syringe lying on the table]. Like that's the air molecules	

[a]Words that are in *bold, italics and underlined* are the speech parts that were accompanied with a gesture

5.6 Enactment of H2: Sensemaking of Entities and Activities at the Particle Level

Soon after this interaction, I provided them with a task sheet containing instructions and conceptual questions to guide their engagement with the simulation. Based on the first instruction that asked them to describe what they noticed about the simulation; they correlated the rate with which they tapped their fist to the position of the wall in the simulation that represented the plunger. Lines 103 and 104 show that they noticed the wall moved inwards the faster they gestured and conversely, moved back as their gesturing slowed down. Both girls spontaneously represented the wall in the simulation with their hands while speaking to each other. They also attended to the movement of the air particles, with Nina repeating her analogy of bouncy balls (Line 107). From this interaction, we can see that describing and using the gesture augmented simulation activated relevant entities and their regular activities for the pair. However, this moment also indicates that they had not yet attended to the causal interactions in the simulation.

The next instruction on the task sheet directed them to attend to the red dots on the wall. These red dots were a visual scaffold directing attention toward particle collision, which is a relevant activity from the perspective of MR. However, our prior analysis showed that some students struggled with recognizing the relevance of the red dots, and thus, we amplified its significance by including a written instruction on the sheet aside from the visual cue.

As Nina read the instructions, Adrianna repeatedly tapped her fist against her palm to move the plunger and observe the red dots (Line 109–110). The combination of visual cues, gesturing, and written instructions appeared to help Adrianna grasp the conceptual connection between the red dots and particle collisions (Lines 112). On Nina's inquiry, Adrianna shared her new learning through pointing gestures and verbal explanations maintaining a shared understanding of the concept (Table 5.2).

Table 5.2 Excerpt of Nina and Adrianna exploring the role of the plunger

Line	Speaker	Transcript	Video capture
		Researcher handed them a sheet with an instruction: "Talk with your partner about all the things you notice"	
103	Nina:	You can tell when it [*control gesture*] goes faster that the thing is moving more and when you go slower, it goes ***back***	
104	Adrianna:	Like the ***plunger***, like, like, pushes forward more	

(continued)

Table 5.2 (continued)

Line	Speaker	Transcript	Video capture
105	Nina	Yeah	
105	Adrianna:	Like, it's fun to play with	
107	Nina:	'Cause it's like bouncy balls	
108	Researcher:	Sure, of course. You might want to try— if you've seen those questions, these are another set	
109	Nina:	[*Reads question on sheet*] Count the number of red dots on the plunger to go up and down by tapping your fist to your palm	
110	Adrianna:	Oh, those are what the red dots are! [*while using directed gesture*]	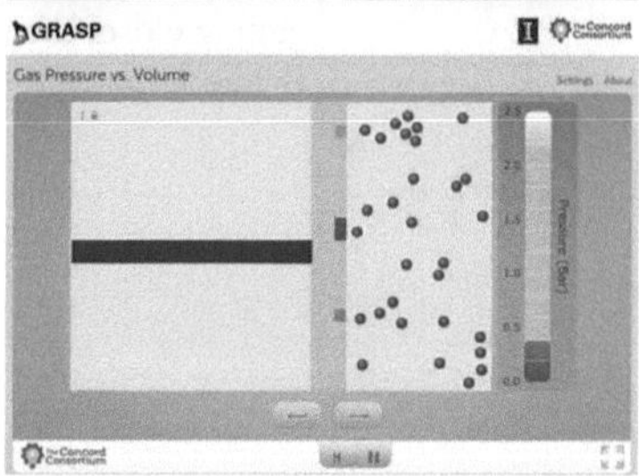
111	Nina:	What?	
112	Adrianna	The ***red dots*** is where the molecules hit it	

Nina then continued reading the guiding question on the sheet (Line 113) while Adrianna controlled the simulation. The instructions asked the students to describe the simulation when the number of red dots were low. The following question led the pair and myself to make several insights in quick succession that linked the speed of bumping fist to palm to the number of red dots on the wall (Nina, line 113), the number of red dots to the pressure indicator on the right of the screen (Adrianna, Line 117), the emergence of the red dots at the point of molecular collision (Nina, Line 118) and the reduction in volume associated with increased rate of molecular movement (Adrianna, Line 119). Their identification of these interactions appeared to help the pair make the conceptual leap that the air molecules hit the wall less frequently when they had more space (Lines 121 and 122) (Table 5.3).

Table 5.3 Excerpt of Nina and Adrianna correlating pressure and volume

Line	Speaker	Transcript	Video Capture
113	Nina:	Yeah. [*starts reading the sheet*] When is the number of red dots low? [*looks up at the screen]* It's low when like you slow down and then it's high when you go faster	
114	Adrianna	The pressure bar	
115	Nina	Yeah	
116	Researcher	The red dots, isn't it? When do the red dots- when are they higher?	
117	Adrianna	Oh, because the pressure— when the pressure is higher—	
118	Nina	The ball like— the molecules like start bouncing off of it—and they keep hitting it yeah	
119	Adrianna	They get closer together, they start getting closer together and… [*while using control gesture*]	 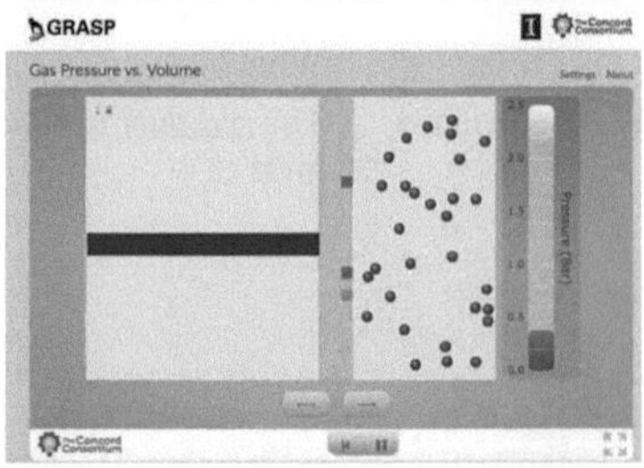
120	Researcher	And when are the number of red dots lower?	
121	Adrianna	When it's **spaced out**. [*spreads hands apart]*	

(continued)

Table 5.3 (continued)

Line	Speaker	Transcript	Video Capture
122	Nina	Because it's a lesser chance of ***hitting it*** [*right palm with fingers touching moves in the direction of molecules hitting on screen*]	

In terms of mechanistic reasoning, this phase focused on describing how entities and activities are linked at this level. Here we see how the collective efforts of written instructions, gesturing, and visual cues supported the pair in making several quick conceptual connections.

5.7 Continued Enactment of H2: Addressing Misconceptions and Reifying the Mechanisms

Although Nina and Adrianna together unpacked relevant activities with the help of the simulation, they had at this point, not yet accounted for the meaning of pressure. Prior research shows that students often conceive of air molecules moving in a background of air (Tytler, 1998). Hence, the following instructional task guided the pair to deepen their understanding of the organization (de Andrade et al., Chap. 4, this book) of air particles by directing the pair to discuss this misconception (Nina, Line 144). In Table 5.4, line 146, Nina entertained this new information and tested it against the activities of wall positioning, molecular collisions, and pressure gauge that the pair had just uncovered.

As Nina described the rising rate of collisions, her description was reified by her controlling the simulation and testing it (line 148). This testing supported the claim about molecules making up the air. Thus, when Nina controlled the simulation and varied the rate of tapping her palm to change the rate of collisions (Lines 146–152), the pair confirmed the statement from the task sheet and resolved any potentially alternative conceptions they might have had about air molecules. During Nina's explanation, the role of the directed gesture took centerstage as Nina varied the rate of her fist tapping her palm to align with her verbal claims allowing both students to deepen their understanding of the relationships across all entities. Additionally, Adrianna's role as her partner cannot be undermined as well for encouraging Nina to gesture and supporting Nina's explanation with her own observations about the rate of molecular collisions (Line 152).

Table 5.4 Nina and Adrianna reason about air molecules in a vacuum

Line	Speaker	Transcript	Video Capture
144	Nina:	Some think that molecules are moving around in the background of air, but scientists think that molecules are made of air— uh, (Adrianna: Making up air.) yeah. There is nothing touching the plunger but the moving molecules	
145	Adrianna:	Oh…. Ooooh!	
146	Nina:	Maybe…maybe that is true, because when like you're pushing it [*the wall*] the molecules get like, they move more. You could see like they're always hitting it	
147	Adrianna:	Can you do it? [*slides device towards Nina*]	
148	Nina:	like when it goes ***like this*** [uses control gesture], it's always hitting it, no matter what. It's just in a bigger place	
149	Researcher:	[inaudible], how many are hitting it?	
150	Nina:	It's like a couple, maybe like when you go like that—	
151	Adrianna:	There's probably like 10 a second. No, no. five a second. And then when it's really close together, then the hits it about a couple of times	
152	Nina:	Like when you go like [*taps palm slowly*] it's about maybe 10 or 5 a second and then when you go like that [*taps palm faster*], it moves and then like a bunch of them are hitting it. So maybe that's true that like the air is the molecules	

5.8 Enactment of H3: Coordinating Relationships Across Scalar Levels

As a final step in MR, students should verify how well the underlying mechanisms align with the target phenomenon. To facilitate this, I demonstrated the syringe to the pair and asked them to explain the events. The pair used a combination of gestures to relate the representations in the simulation to the syringe. Nina represented the space between the plunger and its sealed end, bringing her hand toward the simulation to make it "even with it" (Line 167), thereby linking the simulation to the target phenomenon. Adrianna built on Nina's claim by specifically identifying the simulation's wall as the plunger (Line 168). When I asked to explain why the plunger came back out, Nina gestured with her right hand (line 174) with the closing and opening hand to demonstrate the "stuff" inside pushing the wall back. Notably, she used this same gesture while discussing pushing before (Line 122) when describing the molecules hitting the screen. Similarly, Adrianna expressed the push with her exaggerated moves making a sound effect with her pushing gesture (Line 175) which is quite reminiscent of the greater range of motion and expressions she generally used throughout this interview (Table 5.5).

In this way, by revisiting the target phenomenon at the end of simulation-use, the pair related the representations of simulation and the syringe and coordinated the relevant interactions across scalar level to explain air pressure.

Table 5.5 Excerpt of Nina and Adrianna concluding their exploration

Line	Speaker	Transcript	Video Capture
167	Nina:	I was looking at like from ***<u>sideways</u>*** [*Right hand thumb and index finger are extended to represent space inside the syringe*]. I like made it even with it to like see like when you press it and see	
168	Adrianna:	Yeah, ***<u>this part</u>*** [*the wall in the simulation*] is ***<u>that</u>*** [*points to the plunger*]	
169	Nina:	and then like when you press it you can see like where it goes	
170	Researcher:	Mm-hm, mm-hm. So does that make sense why it's getting— why it gets harder and harder to push the plunger in?	
171	Adrianna:	Yeah	
172	Nina:	Yeah, 'cause there's so much stuff in it that it's like blocking it so you can't press it all the way in to where there's nothing in it	
173	Researcher:	Ok. And when you let go, why is it coming back out?	

(continued)

Table 5.5 (continued)

Line	Speaker	Transcript	Video Capture
174	Nina:	Because of all the stuff in it. Like, it just ***pushes*** it back out	
175	Adrianna:	The pressures that got build up, but '***pewww'***. [*laughs*]	

5.9 Discussion

This chapter examined how directed gesturing in a visually rich, collaborative augmented reality environment facilitates students' MR. Using Krist et al.'s (2019) three epistemic heuristics of MR, I showed how these heuristics were facilitated through gestures, simulation design, and instructional task design. While only one student could control simulation at a time, studying how pairs of students used the simulation helped reveal the types of interactions that are likely to occur in classroom settings and how directed gesturing can enhance MR. The case analysis suggests that directed gesturing integrated seamlessly into Nina and Adrianna's MR processes facilitating their reasoning in a few ways.

The simulation introduced the target phenomenon at the level of air particles, thereby facilitating the first epistemic heuristic of MR and activating students' analogical reasoning. Nina's comparison to bouncy balls offered the first analogical basis for Nina to understand molecular interactions. This observation confirms existing research on science-based computer simulations, which are particularly effective at helping students grasp unobservable entities and activities (Honey & Hilton, 2011).

The pair used both directed gesture and their own spontaneous gestures to enhance their sensemaking and communication of the target mechanism. The directed gesture appeared to help them focus on molecular collisions even when discussing related activities rather than the collisions themselves. From an embodied cognition perspective, this is significant. Their consistent use of the simulation gesture made the molecules hitting the wall particularly prominent in their discourse and thinking. By maintaining this schema in their unconscious explanatory modeling, the pair could build upon this foundation while engaging with other conscious-level semiotic resources (Mathayas et al., 2021).

While making sense of the simulation and coordinating across the simulation and the plunger, Nina and Adrianna used pointing gestures to establish shared intersubjectivity between the two. However, they also used their own spontaneous representational gestures when communicating their insights towards the end of the task. Of note, neither student gestured at the beginning of the interview. This suggests that the directed gesturing through the simulation may have helped broaden their repertoire of representational practices in science further supporting their knowledge building practices (Prain & Tytler, 2012).

Similar to Bachtiar et al.'s (2024) findings that guiding questions promote higher levels of MR, case analysis showed that the questions guided students through distinct phases: describing the simulation (H1), identifying activities (H2), and deepening understanding (H2). The instructional questions were strategically designed based on research and previous implementations, addressing known misconceptions of middle school students while aligning with the heuristics. This structured approach significantly contributed to the students' comprehension of the concepts.

While our prior research focused on how directed gesturing fostered individual students' intuitive and conscious explanatory modeling, this study expanded the scope to explore collaborative use of the gesture-augmented simulation. The collaborative discussions between Nina and Adrianna about conceptual ideas tied to the directed gestures were particularly helpful in fostering their mechanistic reasoning. This finding aligns with other researchers who note that although moments of embodied learning are brief, students grasp scientific principles best when given space to discuss and collectively make sense of their experiences (Danish et al., 2020).

As this single case study focused on one topic and environment, this research demonstrates rich potential for gesture-augmented environments while suggesting broader applications.

5.10 Conclusion

Findings suggest that gesturing can make a profound impact on students' mechanistic reasoning. Both encouraging students to gesture spontaneously and directing them to gesture about relevant activities in the mechanism can enhance their MR at core intuitive and representational levels. However, gesturing itself is not a complete pedagogical tool to support MR. Curricular approaches that support embodied ways of reasoning such as in model based instruction (Berg & Hultén, 2024) or with mixed reality learning environments (Tu et al., 2021) can leverage gestures to enhance students' MR. Care must also be taken to consider individual preferences with gesturing (Özer & Göksun, 2020), multilingual learners (Ma & Jin, 2022), and cultural backgrounds (Özyürek, 2017). Ultimately, while gesturing alone is not a panacea, its strategic incorporation into pedagogy, alongside continuous instructional support and sensitivity to diverse learner needs, holds great promise for deepening students' understanding of complex scientific concepts through enhanced mechanistic reasoning.

References

Amin, T. G. (2015). Conceptual metaphor and the study of conceptual change: Research synthesis and future directions. *International Journal of Science Education, 37*(5–6), 966–991. https://doi.org/10.1080/09500693.2015.1025313

Atit, K., Gagnier, K., & Shipley, T. F. (2015). Student gestures aid penetrative thinking. *Journal of Geoscience Education, 63*(1), 66–72. https://doi.org/10.5408/14-008.1

Bachtiar, R. W., Meulenbroeks, R. F. G., & van Joolingen, W. R. (2022). Mechanistic reasoning in science education: A literature review. *Eurasia Journal of Mathematics, Science and Technology Education, 18*(11), em2178. https://doi.org/10.29333/ejmste/12512

Bachtiar, R. W., Meulenbroeks, R. F. G., & van Joolingen, W. R. (2024). Understanding how student-constructed stop-motion animations promote mechanistic reasoning: A theoretical framework and empirical evidence. *Journal of Research in Science Teaching, 61*(2), 289–318. https://doi.org/10.1002/tea.21891

Barsalou, L. W. (2008). Grounded cognition. *Annual Review of Psychology, 59*(1), 617–645. https://doi.org/10.1146/annurev.psych.59.103006.093639

Belhiah, H. (2013). Gesture as a resource for intersubjectivity in second-language learning situations. *Classroom Discourse, 4*(2), 111–129. https://doi.org/10.1080/19463014.2012.671273

Berg, A., & Hultén, M. (2024). Children's emergent mechanistic reasoning in chemistry: A case study about early primary students' reasoning about the phenomenon of thermal expansion of air. *Chemistry Education Research and Practice, 25*(1), 92–114. https://doi.org/10.1039/D3RP00169E

Bolger, M. S., Kobiela, M., Weinberg, P. J., & Lehrer, R. (2012). Children's mechanistic reasoning. *Cognition and Instruction, 30*(2), 170–206. https://doi.org/10.1080/07370008.2012.661815

Brown, D. E. (2018). Implicit conceptual dynamics and students' explanatory model development in science. In T. G. Amin & O. Levrini (Eds.), *Converging perspective on conceptual change: Mapping an emerging paradigm in the learning sciences* (pp. 105–112). Routledge.

Cheng, M.-F., & Brown, D. E. (2010). Conceptual resources in self-developed explanatory models: The importance of integrating conscious and intuitive knowledge. *International Journal of Science Education, 32*(17), 2367–2392. https://doi.org/10.1080/09500690903575755

Clement, J. J. (2009). *Creative model construction in scientists and students: The role of imagery, analogy, and mental simulation*. Springer.

Clement, J. J. (2013). Roles for explanatory models and analogies in conceptual change. In S. Vosniadou (Ed.), *International handbook of research on conceptual change* (2nd ed., pp. 412–446). Routledge.

Cook, S. W., Yip, T. K., & Goldin-Meadow, S. (2012). Gestures, but not meaningless movements, lighten working memory load when explaining math. *Language and Cognitive Processes, 27*(4), 594–610. https://doi.org/10.1080/01690965.2011.567074

Cook, S. W., Fenn, K. M., Church, R. B., Alibali, M. W., & Kelly, S. D. (2017). The function of gesture in learning and memory. In *Why Gesture?: How the hands function in speaking, thinking and communicating* (Vol. 7, pp. 129–153). John Benjamins Publishing Company.

Crowder, E. M. (1996). Gestures at work in sense-making science talk. *The Journal of the Learning Sciences, 5*(3), 173–208.

Danish, J. A., Enyedy, N., Saleh, A., & Humburg, M. (2020). Learning in embodied activity framework: A sociocultural framework for embodied cognition. *International Journal of Computer-Supported Collaborative Learning, 15*(1), 49–87. https://doi.org/10.1007/s11412-020-09317-3

de Andrade, V., Shwartz, Y., Freire, S., & Baptista, M. (2022). Students' mechanistic reasoning in practice: Enabling functions of drawing, gestures and talk. *Science Education, 106*(1), 199–225. https://doi.org/10.1002/sce.21685

DeSutter, D., & Stieff, M. (2017). Teaching students to think spatially through embodied actions: Design principles for learning environments in science, technology, engineering, and mathematics. *Cognitive Research: Principles and Implications, 2*(1), 22. https://doi.org/10.1186/s41235-016-0039-y

Erb, C. D., Buchanan, D. W., & Sobel, D. M. (2013). Children's developing understanding of the relation between variable causal efficacy and mechanistic complexity. *Cognition, 129*(3), 494–500. https://doi.org/10.1016/j.cognition.2013.08.002

Flood, V. J., & Harrer, B. W. (2023). Kinetically-held questions: Representational gesture post-stroke holds in whole-class interactions in STEM. *Linguistics and Education*, 101164. https://doi.org/10.1016/j.linged.2023.101164

Gentner, D. (1989). The mechanisms of analogical learning. In S. Vosniadou & A. Ortony (Eds.), *Similarity and analogical reasoning* (pp. 199–241). Cambridge University Press.

Givry, D., & Roth, W.-M. (2006). Toward a new conception of conceptions: Interplay of talk, gestures, and structures in the setting. *Journal of Research in Science Teaching, 43*(10), 1086–1109. https://doi.org/10.1002/tea.20139

Glennan, S., Illari, P., & Weber, E. (2022). Six theses on mechanisms and mechanistic science. *Journal for General Philosophy of Science, 53*(2), 143–161. https://doi.org/10.1007/s10838-021-09587-x

Goldin-Meadow, S. (2011). Learning through gesture: Gesture. *Wiley Interdisciplinary Reviews: Cognitive Science, 2*(6), 595–607. https://doi.org/10.1002/wcs.132

Goldin-Meadow, S., Nusbaum, H., Kelly, S. D., & Wagner, S. (2001). Explaining Math: Gesturing lightens the load. *Psychological Science, 12*(6), 516–522. https://doi.org/10.1111/1467-9280.00395

Goldin-Meadow, S., Cook, S. W., & Mitchell, Z. A. (2009). Gesturing gives children new ideas about math. *Psychological Science, 20*(3), 267–272.

Goodwin, C. (2000). Action and Embodiment Within Situated Human Interaction. *Journal of Pragmatics, 32*, 1489–1522.

Hegarty, M. (2004). Mechanical reasoning by mental simulation. *Trends in Cognitive Sciences, 8*(6), 280–285. https://doi.org/10.1016/j.tics.2004.04.001

Hofstadter, D. R. (2001). Analogy as the core of cognition. In D. Gentner, K. J. Holyoak, & B. N. Kokinov (Eds.), *The analogical mind: Perspectives from cognitive science* (pp. 499–538). The MIT Press.

Honey, M. A., & Hilton, M. L. (2011). *Learning science through computer games and simulations*. National Academies Press. https://doi.org/10.17226/13078

Hostetter, A. B., & Alibali, M. W. (2008). Visible embodiment: Gestures as simulated action. *Psychonomic Bulletin & Review, 15*(3), 495–514. https://doi.org/10.3758/PBR.15.3.495

Hostetter, A. B., & Boncoddo, R. (2017). Gestures highlight perceptual-motor representations in thinking. In R. B. Church, M. W. Alibali, & S. D. Kelly (Eds.), *Why Gesture?: How the hands function in speaking, thinking and communicating* (Vol. 7, pp. 155–174). John Benjamins Publishing Company. https://doi.org/10.1075/gs.7.08hos

Johnson, M. (2009). *The body in the mind: The bodily basis of meaning, imagination, and reason* (Paperback ed., [9. print.]). University of Chicago Press.

Jordan, B., & Henderson, A. (1995). Interaction analysis: Foundations and practice. *The Journal of the Learning Sciences, 4*(1), 39–103.

Kapon, S. (2017). Unpacking sensemaking. *Science Education, 101*(1), 165–198. https://doi.org/10.1002/sce.21248

Kita, S., Alibali, M. W., & Chu, M. (2017). How do gestures influence thinking and speaking? The gesture-for-conceptualization hypothesis. *Psychological Review, 124*(3), 245–266. https://doi.org/10.1037/rev0000059

Krist, C., Schwarz, C. V., & Reiser, B. J. (2019). Identifying essential epistemic heuristics for guiding mechanistic reasoning in science learning. *Journal of the Learning Sciences, 28*(2), 160–205. https://doi.org/10.1080/10508406.2018.1510404

Lakoff, G. (1993). The contemporary theory of metaphor. In A. Ortony (Ed.), *Metaphor and thought* (2nd ed., pp. 202–251). Cambridge University Press.

Lindgren, R. (2015). Getting into the Cue: Embracing technology-facilitated body movements as a starting point for learning. In V. R. Lee (Ed.), *Learning technologies and the body: Integration and implementation in formal and informal learning environments* (Vol. 135, pp. 39–54). Routledge.

Lindgren, R., Wallon, R. C., Brown, D. E., Mathayas, N., & Kimball, N. (2016). *"Show me" what you mean: Learning and design implications of eliciting student gesture in student explanations*. International Conference of the Learning Sciences, Singapore.

Lira, M. E., & Stieff, M. (2018). Using gesture analysis to assess students' developing representational competence. In K. L. Daniel (Ed.), *Towards a framework for representational competence in science education* (pp. 205–228). Springer International Publishing. https://doi.org/10.1007/978-3-319-89945-9_10

Ma, S., & Jin, G. (2022). The relationship between different types of co-speech gestures and L2 speech performance. *Frontiers in Psychology, 13*, 941114. https://doi.org/10.3389/fpsyg.2022.941114

Machamer, P., Darden, L., & Craver, C. F. (2000). Thinking about mechanisms. *Philosophy of Science, 67*(1), 1–25. https://doi.org/10.1086/392759

Mathayas, N., Brown, D. E., & Lindgren, R. (2021). "I got to see, and I got to be a part of it": How cued gesturing facilitates middle-school students' explanatory modeling of thermal conduction. *Journal of Research in Science Teaching, 58*(10), 1557–1589. https://doi.org/10.1002/tea.21718

McNeill, D. (1992). *Hand and mind: What gestures reveal about thought.* University of Chicago Press.

Nathan, M. J., & Martinez, C. V. J. (2015). Gesture as model enactment: The role of gesture in mental model construction and inference making when learning from text. *Learning: Research and Practice, 1*(1), 4–37. https://doi.org/10.1080/23735082.2015.1006758

Nathan, M. J., Alibali, M. W., & Church, R. B. (2017). Chapter 13. Making and breaking common ground. How teachers use gesture to foster learning in the classroom. In *Why Gesture?: How the hands function in speaking, thinking and communicating* (Vol. 7, pp. 285–316). John Benjamins Publishing Company. https://benjamins.com/#catalog/books/gs.7.14nat/details

Niebert, K., & Gropengiesser, H. (2015). Understanding starts in the mesocosm: Conceptual metaphor as a framework for external representations in science teaching. *International Journal of Science Education, 37*(5–6), 903–933. https://doi.org/10.1080/09500693.2015.1025310

Niebert, K., Marsch, S., & Treagust, D. F. (2012). Understanding needs embodiment: A theory-guided reanalysis of the role of metaphors and analogies in understanding science. *Science Education, 96*(5), 849–877. https://doi.org/10.1002/sce.21026

Novack, M. A., & Goldin-Meadow, S. (2017). Chapter 17. Understanding gesture as representational action: A functional account of how action and gesture differ with respect to thinking and learning. In R. B. Church, M. W. Alibali, & S. D. Kelly (Eds.), *Why Gesture?: How the hands function in speaking, thinking and communicating* (Vol. 7, pp. 381–396). John Benjamins Publishing Company. https://doi.org/10.1075/gs.7.18nov

Novack, M. A., Congdon, E. L., Hemani-Lopez, N., & Goldin-Meadow, S. (2014). From action to abstraction: Using the hands to learn math. *Psychological Science, 25*(4), 903–910.

Novack, M. A., Congdon, E. L., Wakefield, E. M., & Goldin-Meadow, S. (2018). Gesture's role in reflecting and fostering conceptual change. In T. G. Amin & O. Levrini (Eds.), *Converging perspectives on conceptual change: Mapping an emerging paradigm in the Learning Sciences* (pp. 97–104). Routledge.

Özer, D., & Göksun, T. (2020). Gesture use and processing: A review on individual differences in cognitive resources. *Frontiers in Psychology, 11*, 573555. https://doi.org/10.3389/fpsyg.2020.573555

Özyürek, A. (2017). Function and processing of gesture in the context of language. In R. B. Church, M. W. Alibali, & S. D. Kelly (Eds.), *Why Gesture?: How the hands function in speaking, thinking and communicating* (Vol. 7, pp. 39–58). John Benjamins Publishing Company.

Padalkar, S., & Ramadas, J. (2011). Designed and spontaneous gestures in elementary astronomy education. *International Journal of Science Education, 33*(12), 1703–1739. https://doi.org/10.1080/09500693.2010.520348

Plummer, J. D. (2009). Early elementary students' development of astronomy concepts in the planetarium. *Journal of Research in Science Teaching, 46*(2), 192–209. https://doi.org/10.1002/tea.20280

Pozzer-Ardenghi, L., & Roth, W.-M. (2007). On performing concepts during science lectures. *Science Education, 91*(1), 96–114. https://doi.org/10.1002/sce.20172

Prain, V., & Tytler, R. (2012). Learning through constructing representations in science: A framework of representational construction affordances. *International Journal of Science Education, 34*(17), 2751–2773. https://doi.org/10.1080/09500693.2011.626462

Roth, W.-M. (2001). Gestures: Their role in teaching and learning. *Review of Educational Research, 71*(3), 365–392.

Roth, W.-M., & Lawless, D. (2002). Scientific investigations, metaphorical gestures, and the emergence of abstract scientific concepts. *Learning and Instruction, 12*(3), 285–304. https://doi.org/10.1016/S0959-4752(01)00023-8

Russ, R. S., Scherr, R. E., Hammer, D., & Mikeska, J. (2008). Recognizing mechanistic reasoning in student scientific inquiry: A framework for discourse analysis developed from philosophy of science. *Science Education, 92*(3), 499–525. https://doi.org/10.1002/sce.20264

Schlottmann, A. (1999). Seeing it happen and knowing how it works: How children understand the relation between perceptual causality and underlying mechanism, *15*.

Singer, M. (2017). The function of gesture in mathematical and scientific discourse in the classroom. In R. B. Church, M. W. Alibali, S. D. Kelly, & S. Goldin-Meadow (Eds.), *Why Gesture?: How the hands function in speaking, thinking and communicating* (Vol. 7, pp. 317–329). John Benjamins Publishing Company.

Stephens, A. L., & Clement, J. J. (2010). Documenting the use of expert scientific reasoning processes by high school physics students. *Physical Review Special Topics – Physics Education Research, 6*(2). https://doi.org/10.1103/PhysRevSTPER.6.020122

Tu, X., Humburg, M. A., Danish, J., Davis, B. N., Ryan, Z. D., Vickery, M., Zhou, M., & Mathayas, N. (2021, April 8). Assessing young children's embodied learning of states of matter in a mixed-reality environment. *2021 Annual Meeting of the American Educational Research Association (AERA), held virtually (due to COVID-19).*

Tytler, R. (1998). Children's conceptions of air pressure: Exploring the nature of conceptual change. *International Journal of Science Education, 20*(8), 929–958. https://doi.org/10.1080/0950069980200803

Ünsal, Z., Jakobson, B., Wickman, P.-O., & Molander, B.-O. (2018). Gesticulating science: Emergent bilingual students' use of gestures. *Journal of Research in Science Teaching, 55*(1), 121–144. https://doi.org/10.1002/tea.21415

Chapter 6
Design Principles for Developing and Analyzing Formative Assessments to Support Mechanistic Reasoning

Keenan Noyes, Clare G. -C. Franovic, and Melanie M. Cooper

6.1 Introduction

In STEM education, there is a focus on understanding mechanisms and promoting mechanistic reasoning as a thinking strategy. In this chapter, we discuss several principles of formative assessment design and evaluation as a means by which instructors and researchers can support students to engage in mechanistic reasoning (MR), particularly for students who have not been exposed to the idea that the mechanisms by which phenomena occur are an important part of learning science. There are a number of frameworks that have been proposed to characterize MR in biology, (van Mil et al., 2013) chemistry, (Moreira et al., 2018) and in science broadly, (Krist et al., 2019; Russ et al., 2008) all of which generally define MR as thinking about/predicting/explaining how and why phenomena occur. While a general understanding is useful for communicating the importance of MR, adhering to a specific framework is necessary for detailed and intentional design of tasks that help students to engage in this type of reasoning (i.e., the focus of this chapter). The work we discuss here is dependent on our understanding of MR and formative assessments; as such, we focus on these pillars and relevant frameworks in the introduction.

K. Noyes
Department of Physiology, Michigan State University, East Lansing, MI, USA

C. G. -C. Franovic
Human Biology Program, Michigan State University, East Lansing, MI, USA

M. M. Cooper (✉)
Department of Chemistry, Michigan State University, East Lansing, MI, USA
e-mail: mmc@msu.edu

N. Graulich et al. (eds.), *Exploring Mechanistic Reasoning in Science Education*, Contributions from Science Education Research 17,
https://doi.org/10.1007/978-3-032-10246-1_6

6.2 How We Define Mechanistic Reasoning

In our work, we have largely aligned with the Krist et al. (2019) framework, which outlines MR as a process involving three steps: (1) considering the "level below", (2) identifying important factors at that level, and (3) linking the lower-level factors to the target phenomenon (Fig. 6.1). While other frameworks (Caspari et al., 2018; Moreira et al., 2018; Russ et al., 2008) are certainly applicable and important for thinking about MR, we have found the simplicity of the Krist et al. framework to be the most useful for our research.

The mechanisms that underlie a range of phenomena are highly context specific, and as noted, numerous authors have defined specific schemes for the analysis of MR. These schemes are often aligned to a particular phenomenon and have sometimes distinguished between *mechanistic* and *causal mechanistic* reasoning. For example, in our early work on mechanistic reasoning in the context of acid base reactions, we defined the scalar level below the phenomenon as involving the rearrangement of electrons, and differentiated between explanations that were mechanistic—that is those that invoke the Lewis acid base mechanism, and those that were causal mechanistic that discussed not only how the electrons rearranged but also why they moved from base to acid (Cooper et al., 2016). However, for the sake of simplicity and because we consider a number of different phenomena, we will not make this distinction in this chapter despite the fact that we have done this in the past. (Becker et al., 2016; Cooper et al., 2016).

6.2.1 *Role of Formative Assessments in Learning*

Assessments play an important role in supporting learning in the classroom. One role is as a source of evidence of students' thinking (Mislevy et al., 2003; Pellegrino et al., 2001). That is, students' responses to assessment items provide evidence of what students know and can do. This information is critical for the instructor to reflect on instructional decisions that best support student learning.

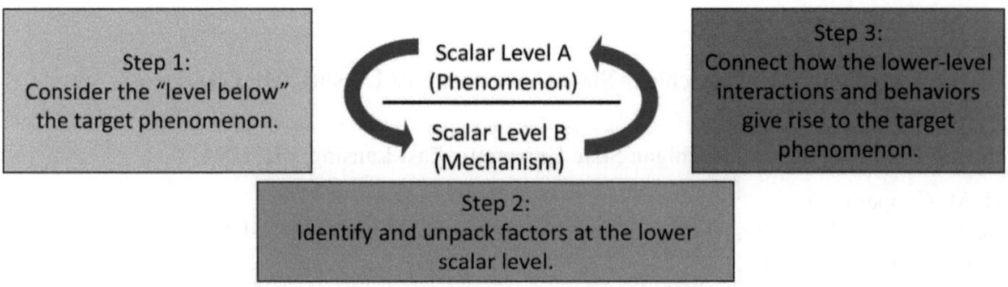

Fig. 6.1 An adaptation of the Krist et al. framework guiding MR (Krist et al., 2019)

Second, formative assessments provide students with an opportunity to apply their knowledge and deepen their understanding. Modern theories of learning suggest that students' knowledge is often initially fragmented and highly dependent on context (diSessa, 2014; Hammer, 2000; Hammer et al., 2005). When students approach a task, the context activates particular components or knowledge elements within their minds (this is what Hammer (Hammer, 2000; Hammer et al., 2005) refers to as "resources") that they use to reason through the problem. These resources may or may not be productive in given situations, but the resources themselves are not inherently correct or incorrect. From this perspective, teaching and learning is about helping students to (1) build stronger connections between relevant resources and (2) use the appropriate resources across different contexts. The building of these connections can only be done by the student (a central tenet of constructivism, which states that knowledge is constructed in the mind of the learner (Bodner, 1986)), so formative assessments are important for providing opportunities to make these connections. This process can be time-consuming and requires considerable effort on the part of the student. Therefore, these kinds of activities are best done in the context of formative assessments where there are opportunities to make mistakes, receive feedback, and recover, rather than the context of high stakes summative assessments (e.g., exams). This paper will focus on the development and use of these formative assessments (also referred to as "tasks").

Finally, assessments are an important tool for messaging to students what is important in both the classroom and in the discipline (Shepard, 2000). Assessments represent what instructors want their students to know and be able to do at the end of the learning process. Assessments are also how we typically give students credit in the course and therefore play an important role in determining if the student will pass the class. This is especially important for introductory courses such as chemistry, which are critical to retention and success in STEM.

While there are many possible approaches to the inclusion and design of formative assessments that have the potential to help students connect and use their knowledge, we propose that mechanistic reasoning should be central to many of the learning activities and tasks that we ask students to complete. Incorporating MR tasks into instruction and assessment can provide instructors with key information about how students learn to explain and predict phenomena, while at the same time providing students with opportunities to develop their own mechanistic understanding of important chemistry content through emphasizing the importance of this type of reasoning. However, effectively designing formative assessments to support MR and evaluating responses for student engagement in MR is not trivial. One common thread across the development of the formative assessments we discuss is the importance of an iterative design process. Even the best laid plans will still run into unforeseen issues when implemented. When we analyze the results of an assessment, we may find that minor aspects of the assessment that were once thought to not be an issue suddenly have ballooned into major issues. Since this is inevitable, how then do we mitigate these issues?

6.2.2 *Iterative Design*

Many of our discussion points in this paper emerge as a result of our modification of evidence-centered design (ECD) as outlined by Mislevy et al. (2003) ECD is an iterative process that aims to create an optimized assessment task via a process that involves (1) deciding what it is that students need to know and how they need to know it, (2) determining what evidence would convince the instructor or research team that students had attained this goal, and then (3) designing a task that has the potential to elicit that evidence. In our work we have acknowledged that ECD is a cyclical process, so that once the task is designed it is (4) administered to students, (5) the student responses are analyzed, and then (6) the task is refined based on student responses. For example, if very few students are including an expected idea, the task may need to be modified to more effectively activate the appropriate resources.

A final, optimized task is one that is balanced with the support students need so that they understand what is expected of them, but not so much that they can answer the task without meaningful effort. That is, the designer needs to find the "sweet spot" with regard to scaffolding in order to achieve the intended outcome. Scaffolds, or techniques to support students as they attempt a task, are varied in nature and can be implemented in several ways, the most successful of which are typically determined through this process of iterative design.

How then should formative assessments be designed to support mechanistic reasoning? In this paper, we provide general principles, related to formative assessments, that ultimately support students to develop and refine the practice of mechanistic reasoning. The principles we highlight emerged through work from the Cooper group, and from the development of the general and organic chemistry CLUE curricula (Cooper et al., 2019; Cooper & Klymkowsky, 2013). We highlight specific findings from these studies in this paper to provide evidence and context for these principles.

6.3 Principles of Formative Assessments for Mechanistic Reasoning

Below, we highlight principles (Fig. 6.2) related to formative assessments that support mechanistic reasoning. These principles include:

- Scaffolding techniques to elicit mechanistic reasoning (Reiser & Tabak, 2014)
  - Breaking up the question into multiple parts (Becker et al., 2016; Cooper et al., 2016)
  - Guiding students to think about the underlying entities (Krist et al., 2019)
- Capturing mechanistic components in a coding scheme (Mislevy et al., 2003)
- Grading for effort and thoroughness (Bowen et al., 2022)

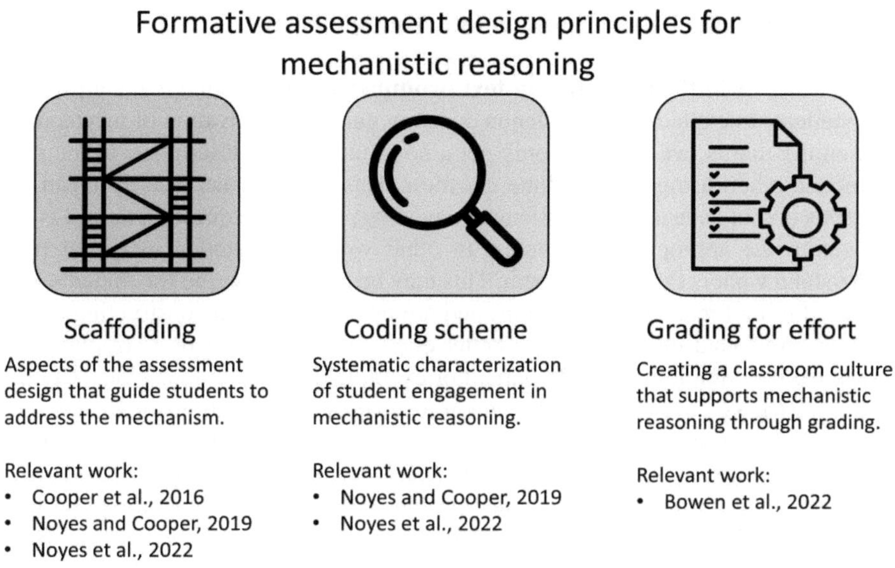

Fig. 6.2 Summary of the design principles highlighted in this paper and related studies conducted by the Cooper group

Throughout, we will discuss several projects studying MR in the context of acid/base reactions, (Cooper et al., 2016; Crandell et al., 2019, 2020) London dispersion forces (LDFs), (Becker et al., 2016; Noyes & Cooper, 2019) and protein-ligand binding. (Franovic et al., 2023; Noyes et al., 2022).

6.3.1 Scaffolding Techniques (Modified ECD # 3, 4, 5, 6)

When we think about formative assessments, scaffolding plays an important role. By scaffolding, we refer to aspects of the assessments that help the student to complete the task (Wood et al., 1976). Techniques such as reducing degrees of freedom or marking critical features can help guide students to activate the appropriate resources and strengthen those connections in their mind. Eventually, these supports can be removed or faded once those are developed enough to stand on their own. While scaffolding was originally described in the context of tutoring or interactions between the instructor and learner, these principles can also be embedded in formative assessments (Sadler, 1989; Shepard, 2005).

Recall that using the ECD framework involves analyzing student responses followed by modifying the task. This modification may require either adding, removing, or refining scaffolds based on the responses collected, with the ultimate aim of achieving optimal balance. In the following sections, we discuss studies that taught us two major scaffolding techniques which we have since relied on for subsequent assessment items in our research and instruction.

6.3.1.1 Breaking Up the Question Into Multiple Parts

Separating "what" and "why" in text prompts: We have found that simply asking students to explain a phenomenon does not guarantee activation of mechanistic reasoning; that is, we typically only get a non-mechanistic description of the phenomenon when using a standalone question. This does not necessarily mean the students do not have a mechanistic understanding; rather, the question design is not activating the appropriate resources. In other words, the student may not fully understand what is expected of them. This may be particularly true for students who are used to multiple choice questions and are not familiar with constructing explanations. We found that one way to better elicit causal mechanistic reasoning is to break the question into multiple parts. For example, in our previous work (Cooper et al., 2016), we developed a task to elicit mechanistic explanations of what occurs during a simple acid-base reaction (Fig. 6.3). However, the initial attempt at eliciting this type of reasoning tended to yield a *description* of the reaction. That is, students would describe that acids donate protons (H^+) to bases which accept protons (H^+) without discussing *how* or *why* the reaction occurs (i.e., the mechanism). Such an explanation would require the students to step down a scalar level to address the role of polarity and partial charges in leading the negatively charged electrons on the oxygen of water molecule to "attack" (form a covalent bond) the hydrogen on HCl. Explanations including these ideas were observed with a significantly greater frequency when the students were asked in two separate questions to first explain what happens and *then* explain why.

This decision to break the question into multiple parts encouraged students to address both what happened (a description) and how/why it occurred (the mechanism). By first asking for a description of what was happening, the goal was that students would then understand that the "why" prompt required them to go further

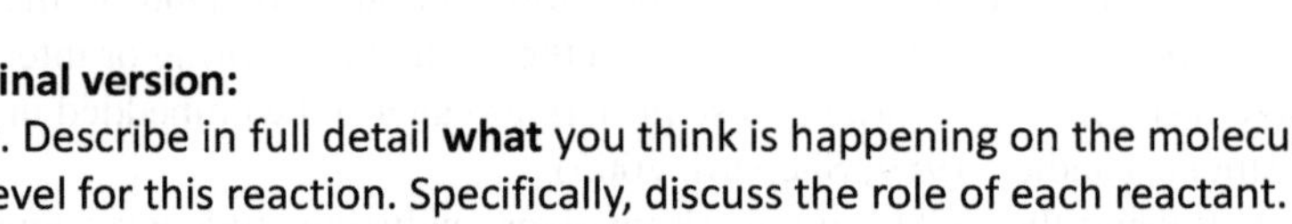

Initial version:
Please explain your reasoning for what you think is happening at the molecular level for this reaction.

Final version:
a. Describe in full detail **what** you think is happening on the molecular level for this reaction. Specifically, discuss the role of each reactant.

b. Using a molecular level explanation, please explain **why** this reaction occurs. Specifically, why the reactants form the products shown

Fig. 6.3 Adapted figure from Cooper et al. (2016). Prompts edited to only show the sections that show the modifications that produced change in the text

and provide explicit reasoning about the phenomenon of proton transfer. These decisions were key in the development of future tasks/research in which we have separated out the "what" and the "why" (Crandell et al., 2020; Noyes et al., 2022).

Targeting the sequence of events in drawing prompts: For some phenomena, a mechanistic explanation must include a temporal account of the behavior of the relevant entities (Krist et al., 2019). This is true for chemical phenomena like dissolving solutes in solvents or the origins of LDFs. However, it can be difficult to elicit the temporal aspect of a phenomena when using drawings as the source of evidence of student thinking. The key is to cue students into considering the sequence of events that occur.

One way to do this is to provide students with a task that makes the sequence explicit. This can be achieved by asking students to draw or explain what is happening at particular stages that lead to the phenomenon. This approach was a key part in our work exploring students' mechanistic reasoning about the origin of LDFs (Becker et al., 2016; Noyes & Cooper, 2019). In this work, we asked students to draw the process by which two neutral atoms would attract one another (via LDFs). Ideally, students should draw the distortion of the electron cloud of one atom (instantaneous dipole) which leads to partial charges on either end of the atom. This instantaneous dipole can then cause attractions and repulsions with the electron cloud of a neighboring atom, distorting its electron cloud and inducing a dipole in that atom. We designed a formative assessment to explore how students draw (and explain) this intermolecular force. While about 1/3 of the students in the sample provided a causal mechanistic explanation for this attraction, virtually none of students in the sample provided a causal mechanistic *drawing*. This would suggest that a meaningful portion of the students had a causal mechanistic understanding of the phenomenon, but the drawing prompt was not eliciting that knowledge.

In the follow up study, we expanded the drawing task from a single box to three boxes. By providing multiple boxes, we hoped to message to the students that we wanted them to draw a process rather than a static image (Fig. 6.4). This modification significantly changed the student responses, showing an increase in causal mechanistic drawings from 2% to 55% of the students (Noyes & Cooper, 2019).

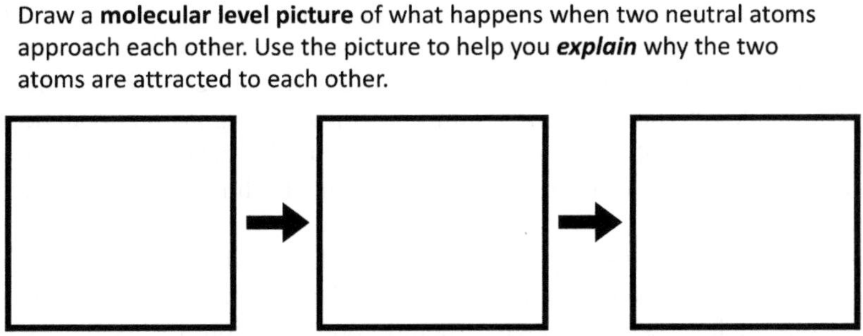

Fig. 6.4 An adapted version of the LDF drawing prompt with the drawing area subdivided into multiple portions to highlight the mechanistic process we hoped to elicit (Noyes & Cooper, 2019)

This approach was also key for the development of the Solvation Representation Assessment (SRA), an instrument to explore student understanding of the mechanism by which a solute dissolves in solution. (Judd Jr., 2019) In the initial version of the SRA, although the vast majority of students were able to draw the final solvated product, few students drew or discussed any kind of mechanism for how this occurred. By adding the scaffolded three boxes (undissolved, dissolving, and dissolved) to the final version, we were able to better prompt students to discuss and visualize the process of solvation. The multiple boxes were key for this assessment to explore students' understanding of the mechanism, as the "dissolving" box provides some support and guidance for students to draw the simultaneous forming and breaking of interactions that occurs during solvation.

Most students have multiple resources that can be activated when responding to a particular task, including both conceptual ideas about the phenomenon at hand as well as ideas about what is expected of them by the prompt/instructor/course. Effectively cueing the students to explain phenomena at a certain depth is not trivial; however, breaking up the question into parts sends the message that something different (perhaps deeper or more explanatory) should be included in each part. As a result, we are provided with more convincing evidence of what students know and can do regarding these phenomena and can, therefore, make better informed decisions about future instructional practices. This simple modification has had a profound impact on the design of questions in our chemistry courses.

6.3.1.2 Guiding Students to Think About the Underlying Entities

To reason mechanistically, students must connect the appropriate underlying entities to the phenomenon of interest. Helping students consider the underlying entities can be quite difficult, especially if there are many factors, entities, or multiple scalar levels the students must traverse. The challenge is finding ways to gently or subtly activate students' knowledge. If the task is over-prompting students and forcing them to use resources they would otherwise see as irrelevant to the task, (1) the connections they make between ideas may not be genuine, (2) the data from the assessments may not be an accurate picture of what the students actually understand, and (3) students may attempt to insert the idea inappropriately and become confused. For example, are students discussing electrons because they believe they are truly important to the phenomenon at hand or because they were explicitly instructed to discuss those ideas? This difference is important for making instructional decisions based on the results of the formative assessment. We have taken a variety of approaches in our prior work to encourage students to think about the underlying entities.

For example, in preliminary development of the acid/base reaction in Cooper et al. (2016) (discussed above), we found that students were more likely to go down to the electronic level when a Lewis structure representation was used instead of a condensed structure. The subtle cue of the presence of covalent bonds and lone pairs encouraged students to go to the appropriate scalar level to provide a mechanistic explanation for this phenomenon. Even so, it was the later addition of the "what" and "why" prompts that produced the highest proportion of mechanistic responses.

As phenomena increase in complexity, the design of question prompts becomes even more important. A more recent study of ours involved the design of a task that sought to elicit causal mechanistic explanations about protein-ligand binding, a phenomenon that typically occurs as a result of electrostatic interactions and is discussed in both chemical and biological contexts (Noyes et al., 2022). The final optimized prompt emerged after three major iterations, a process that gave us two insightful approaches for better guiding students to activate relevant ideas.

The first approach involved applying suggestions from Graulich and co-workers, (2020; Graulich & Schween, 2018) where contrasting cases were used to cue students into the relevant components or scalar level at which the mechanism occurs. Using this approach in the context of protein ligand binding, we asked students to compare two protein binding sites and choose which of those sites would more likely bind a ligand. Figure 6.5 shows the two binding sites as they were presented to the students (in the final version of this task), which differ by a single amino acid: a non-polar alanine side chain versus a polar asparagine side chain. The image shows all of the atoms in these different functional groups, providing a subtle message about the scalar level we want the students to discuss (that is, the atomic/functional group level, since the properties of these atoms/groups causes the difference in binding).

The second aspect of this prompt which proved to significantly impact student responses, was using a positively charged ion as opposed to a neutral glucose molecule as the ligand. Electrostatic interactions are the root cause of binding, and by using a positively charged Mg^{2+} ion as the ligand, we could subtly cue the students to think about charge, ideally activating ideas like Coulomb's law which states that opposite charges attract. Thus, students could leverage these simpler ideas to explain something complex, unfamiliar, and perhaps discussed differently in a biology course.

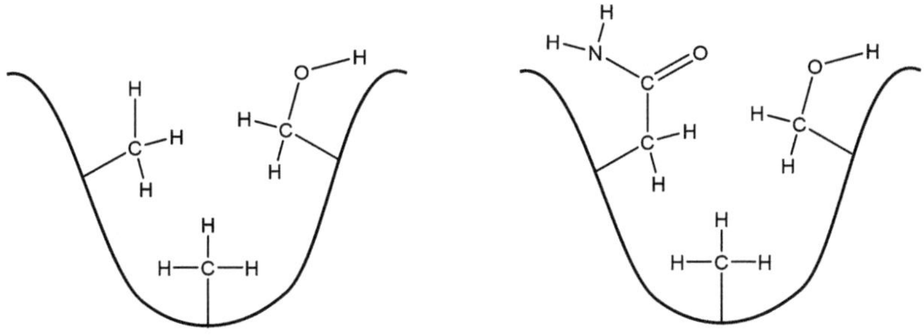

Protein binding site 1 Protein binding site 2

Fig. 6.5 An adapted image of the protein binding sites from the protein-ligand binding task (Noyes et al., 2022)

6.3.2 Capturing Mechanistic Components in the Coding Scheme (Modified ECD # 2, 5)

Once the task has been initially developed, a rubric or coding scheme must be developed to characterize how the student is using causal mechanistic reasoning in the context of that phenomenon. Although ECD specifies the evidence that the task is intended to elicit, and these evidence statements can form the basis of the initial coding scheme, as we have noted, the task often does not meet the intentions of the designers; after all, the coding scheme must capture the actual responses of students, not just the intended responses. This is what provides the evidence (or lack thereof) of student understanding that can then guide task iteration and future instruction. The development of the coding scheme should refer back to the evidence statements (defined during ECD), while also leveraging the MR framework used to define what "counts" as mechanistic. In this way, we can assess whether students are engaging in this type of reasoning and using appropriate resources to do so.

For example, consider the LDF coding scheme (Fig. 6.6). In that task, our goal was for students to explain why neutral atoms would attract one another (the phenomenon). To reflect on what is necessary for this mechanistic explanation, we need to define the evidence statements, or the fine-grained ideas that students should include in order for us to believe they understand those concepts. For this phenomenon, and according to the Krist framework (2019), we need to first think about what the relevant entities are at the lower scalar level. In this case, that would be the subatomic particles, the electrons and protons. Next, what are the relevant properties and behaviors of those entities that someone would need to discuss in order to

	Description	LDF example	Protein-ligand binding example
Non-mechanistic	Describes "what" happens without addressing "why" or "how"	*"Neutral atoms attract because of the London dispersion force"*	*"The magnesium binds to the protein because it fits better"*
Partially mechanistic	Identifies some (but not all) of the underlying entities/properties and does not link back to the phenomenon	*"They attract because the positive end of one atom attracts the negative end of the other"*	*"Mg^{2+} binds to the protein because it is positively charged and binds to the negative amino acid in the binding site"*
Fully mechanistic	Identifies all relevant underlying entities/properties and links those behaviors to the observed phenomenon	*"The electrons may be unevenly distributed around the nucleus leading to partially positive and negative ends of the neutral atoms. The oppositely charged ends can then attract"*	*"The protein I chose has more negatively charged oxygen atoms, so it will more strongly attract the positively charged magnesium ion leading them to bind together"*

Fig. 6.6 Basic components of a coding scheme designed to characterize engagement in causal mechanistic reasoning and hypothetical students' responses to the LDF and protein-ligand binding tasks

explain why this phenomenon is possible? They would need to identify that the electrons and protons have opposite charges. They would also need to address that the electrons are not static and can move around and be unevenly distributed around the nucleus. There are certainly other properties/behaviors of these entities that exist (e.g., mass) but are not relevant in this particular context. Finally, how do students need to link the properties and behaviors of the underlying entities to explain the phenomenon? In this context, students would need discuss how the movement of the charged electrons leads to the formation of temporary and induced dipoles, giving partial charges to otherwise neutral atoms. This leads to the attraction of the oppositely charged ends of the dipoles.

It is also important to examine student responses to reflect on what is actually reasonable for the students to include in their explanation. For example, in the protein-ligand binding coding scheme, we, as chemists, initially believed that students would go down not just to the level of the atom, but also unpack other relevant characteristics of the atom, like the electronegativities of the atoms in each functional group (Noyes et al., 2022). It quickly became clear as we familiarized ourselves with the students' responses, especially those from biology, that unpacking those aspects of the underlying entities was unreasonable, and therefore we did not capture that idea in our coding scheme.

In our recent projects, we have moved to condense our causal mechanistic codes into fewer bins, distinguishing between those who have not addressed the cause or included mechanistic components, those who have provided some of the components, and those who have provided all of them. This has helped us make sense and communicate our results to a broader audience, especially those who have an interest in MR and less so in the conceptual details that are relevant to an isolated phenomenon. For example, in analyzing the protein-ligand binding responses, we identified three important characteristics of responses that provide a fully causal mechanistic explanation. They needed to discuss that (1) oppositely charged species attract, (2) some of the atoms in the site are partially negatively charged, and (3) whichever binding site has the most negatively charged side chains will attract Mg^{2+} (Noyes et al., 2022). These codes can provide instructors with important aspects of protein-ligand binding instruction, but collapsing the bins into broader categories (having no components, all components, or something in between) can tell us something broadly about how they are engaging in causal mechanistic reasoning. We have used a similar approach, collapsing broader bins together, in the LDF work and in the ongoing work in our group (Houchlei et al., 2023; Noyes & Cooper, 2019).

6.3.3 *Grading for Effort and Thoroughness*

Formative assessments are important tools that can provide students with the opportunity to practice their ability to reason mechanistically. However, students must meaningfully engage with the assessments for them to be effective. How then do we build a classroom culture that fosters this type of behavior? Course credit is one important avenue. On one hand, if no credit is given (i.e., the activities are effectively optional), students may not do the assignments. However, if too much credit is given, especially if credit is only given for correctness, students may use outside resources to ensure they get correct answer (Uminski et al., 2023). In fact, a recent study (Glass & Kang, 2022) has shown that the percentage of students who do not benefit from homework assignments because they seek outside help (and copy and paste answers) has risen from around 15% in 2007 to almost 60% in 2018. A balance of incentives and encouragement must be provided for the students to meaningfully reflect on their knowledge of learning these mechanisms in these activities.

In CLUE and OCLUE (Cooper et al., 2019; Cooper & Klymkowsky, 2013), the transformed general and organic chemistry courses in which we emphasize the importance of mechanisms, about *half* of the students' overall grade came from formative assessments. That is, a meaningful portion of students' grades came from engaging in the process of learning rather than being solely determined by high-stakes exams. In general, students received full credit if they put forth good effort on these formative assignments. That is, we did not explicitly grade their responses for correctness. For the in-class clicker questions, students received full credit for answering and not based on if they selected the correct answer. On the homework activities, students received full credit for responding to the multiple-choice, open-ended, and drawing prompts. With this online homework system, beSocratic (Bryfczynski, 2012), we could scan through large numbers of students' responses quickly to ensure that students were truly attempting the activity (e.g., not entering random letters). During the recitations (50-minute discussion sections with ~30 students), students were asked to complete recitation activities with their peers in small groups. As long as students stayed on task and meaningfully engaged with the questions on the activity, they received full credit.

Despite not grading for correctness, the vast majority of students still appeared to give a good faith effort. For example, we administered the LDF task twice during the second semester of general chemistry, once in a homework activity and once on an exam. Despite the difference in assessment environment, we observed no significant difference in the percentages of students using causal mechanistic reasoning (Noyes & Cooper, 2019). Furthermore, the acid/base and protein-ligand binding activities were administered in a low-stakes out-of-class environment, and yet many students still provided causal mechanistic responses (Cooper et al., 2016; Franovic et al., 2023). This all suggests that not only does grading for completion alleviate a good deal of grading responsibility–a practical decision for large-enrollment courses–but students still take the activities seriously.

Although some might worry that by not grading students' responses and not providing individual feedback, we may be inadvertently promulgating student misunderstandings and errors. We attempt to mitigate this by providing class level feedback in the next class session. Representative (anonymous) student responses are chosen for class discussion, to help students understand what a complete appropriate response looks like. Students can see what the range of responses looks like and understand where their response lies. Recently we have used a Generative AI chatbot to provide class level feedback (Cooper & Klymkowsky, 2024) which is also shown to the class. For example, the feedback typically states what percent of the class used mechanistic reasoning, what concepts student used in their responses, and what common mistakes were present, also with corresponding percentages. We also note that we have a rather large corpus of research comparing students in transformed courses that focus on mechanistic reasoning with students in traditional courses, and we find that the students in transformed courses tend to have significantly better learning outcomes (Matz et al., 2018; Underwood et al., 2023). We have also begun to experiment with Generative AI chatbots that can provide individual student feedback (Cooper & Klymkowsky, 2024) as discussed in the future work section.

This type of environment has an impact on students. Bowen et al. (Bowen et al., 2022) studied how students enrolled in traditional and OCLUE organic chemistry courses perceived what types of activities they were assessed on, and how they were expected to think about reaction mechanisms. They found that the majority of OCLUE students felt that they were being asked to engage in ways of thinking which aligned with mechanistic reasoning. In contrast, traditional students were far more likely to report being assessed on their rote knowledge, specifically by memorizing class materials. This finding highlights the role of assessments in determining what is important and valued in the class. If we want students to know that explaining/predicting mechanisms is important, we need to include these ideas on assessments. However, developing mechanistic understanding is difficult, so we need to provide opportunities for students to practice these abilities on formative assessments so they can work through these tough ideas and build these connections. By giving students credit on these assessments in a way that allows them to be messy, we can build a safe environment—one that does not penalize them for initial mistakes and better supports their learning.

6.4 Summary

Including mechanistic reasoning in teaching and assessment can be fraught with problems. It is highly unlikely that students will respond to a task with a mechanistic explanation or prediction unless that task is specifically designed to elicit that kind of thinking. Many instructors are disappointed when they ask students to explain why something happens, because few students inherently understand what is being asked of them. Therefore, we propose that an intentional, iterative, scaffolded design process must be undertaken, to (1) convey to the student what is needed for an appropriate answer, and (2) determine just what evidence is required for a mechanistic explanation. We have outlined both the use of our modified approach to evidence-centered design, in which we emphasize not only the delineation of what evidence the designed task should elicit, but also the iterative process required to ensure that the task functions as intended.

The iterative nature of this process is one of the most important design factors in developing any task, otherwise a researcher or instructor cannot conclude with any certainty just what a student does or does not know about the phenomenon at hand. For example, in developing the protein-ligand binding task, we initially emphasized the role of electrostatics, asking the students to identify charges on the protein binding site and ligand (Noyes et al., 2022). In response, students discussed charge (a component of a fully mechanistic response) in their explanation of the binding, but it was not clear if the student was using electrostatics to explain binding or if they were including electrostatics because we had included so much prompting about it. We responded by pulling back the scaffolding significantly removing all mention of charge in the task, which then resulted in a majority of responses that did not include charge. In the third attempt, we included a more indirect signposting about charge, swapping the ligand from a glucose molecule to a positively charged magnesium ion (Mg^{2+}). This shift activated the electrostatic ideas students held enough that those who saw it relevant to the phenomenon leveraged these ideas in their explanation, while those who did not did not include it. While these ideas are certainly important for the design of formative tasks used in instruction, researchers also need to carefully adjudicate just what level of prompting is likely to elicit the evidence they are seeking. It is not a reasonable assumption that students cannot engage in mechanistic reasoning if the prompt does not give them sufficient cues to indicate that is what is required.

6.5 Future Directions

There is still much we do not know about how to support mechanistic reasoning. For example, there are few studies on the effect of fading the scaffolded supports (McNeill et al., 2006; van de Pol et al., 2010), or how to help students navigate through unfamiliar scenarios (Houchlei et al., 2021), or across disciplines (Schwarz et al., 2020). Much of this work has been stymied by the practicality of having large numbers of students construct and use mechanistic explanations and predictions, which has been limited by the time it takes to score or code such activities. In our transformed courses students practice constructing explanations on models on an almost daily basis, but as discussed, these tasks are formative and are not graded. Researchers may have the resources to code large numbers of student responses, but in day-to-day teaching this is typically not feasible. In recent years several research groups have experimented with machine learning systems that can automatically score student responses and there are several chapters in this book devoted to this approach, but even these systems usually take many coded examples to train the models. For example, in our LDF work with the Automated Assessment of Constructed Response system (AACR), over 700 student responses had to be coded before the model could achieve an acceptable agreement with human coding (Noyes et al., 2020). However, it may well be that the advent of generative AI systems that use Large Language Models will decrease the effort needed to automatically grade and give feedback to both instructors and students. In our preliminary work with Retrieval Augmented Generative (RAG) AI chatbots, we have found that such systems need far less training and are often able to reproduce human coding with only the codebook and a few (N = 10) examples (Cooper & Klymkowsky, 2024).

The use of such Generative AI systems may also solve the problem of providing students with meaningful feedback, although again there is a great deal of work to be done in this area. Although feedback is said to be a crucial component of the learning process (Hattie & Timperley, 2007), the type of feedback is more important. For example, it has been shown that simply providing corrective feedback is often ignored by students (Winstone et al., 2017), and even when students receive such feedback there it may promote the "illusion of learning" (Messeri & Crockett, 2024). Feedback that engages the learner is a more metacognitive dialog has been shown to be more effective than simply providing information about what is wrong with the response and how to correct it (Jonsson, 2013). We have developed prototypes of such systems (Cooper & Klymkowsky, 2024) and plan to implement them in the near future. Ideally, such systems will allow infusion of mechanistic reasoning, into even the largest scale courses.

In the future, we believe that mechanistic reasoning will play a far more prominent role in learning, and the research outlined in this book will provide a solid foundation on which to build this future.

References

Becker, N., Noyes, K., & Cooper, M. M. (2016). Characterizing students' mechanistic reasoning about London dispersion forces. *Journal of Chemical Education, 93*(10), 1713–1724. https://doi.org/10.1021/acs.jchemed.6b00298

Bodner, G. M. (1986). Constructivism: A theory of knowledge. *Journal of Chemical Education, 63*(10), 873. https://doi.org/10.1021/ed063p873

Bowen, R. S., Flaherty, A. A., & Cooper, M. M. (2022). Investigating student perceptions of transformational intent and classroom culture in organic chemistry courses. *Chemistry Education Research and Practice, 23*(3), 560–581. https://doi.org/10.1039/D2RP00010E

Bryfczynski, S. (2012). BeSocratic: An intelligent tutoring system for the recognition, evaluation, and analysis of free-form student input [Ph.D. Dissertation]. Clemson University.

Caspari, I., Kranz, D., & Graulich, N. (2018). Resolving the complexity of organic chemistry students' reasoning through the lens of a mechanistic framework. *Chemistry Education Research and Practice, 19*(4), 1117–1141. https://doi.org/10.1039/C8RP00131F

Cooper, M. M., & Klymkowsky, M. (2013). Chemistry, life, the universe, and everything: A new approach to general chemistry, and a model for curriculum reform. *Journal of Chemical Education, 90*(9), 1116–1122. https://doi.org/10.1021/ed300456y

Cooper, M. M., & Klymkowsky, M. W. (2024). Let us not squander the affordances of LLMs for the sake of expedience: Using retrieval augmented generative AI Chatbots to support and evaluate student reasoning. *Journal of Chemical Education*. https://doi.org/10.1021/acs.jchemed.4c00765

Cooper, M. M., Kouyoumdjian, H., & Underwood, S. M. (2016). Investigating students' reasoning about acid–base reactions. *Journal of Chemical Education, 93*(10), 1703–1712. https://doi.org/10.1021/acs.jchemed.6b00417

Cooper, M. M., Stowe, R. L., Crandell, O. M., & Klymkowsky, M. W. (2019). Organic chemistry, life, the universe and everything (OCLUE): A transformed organic chemistry cur-

riculum. *Journal of Chemical Education, 96*(9), 1858–1872. https://doi.org/10.1021/acs.jchemed.9b00401
Crandell, O. M., Kouyoumdjian, H., Underwood, S. M., & Cooper, M. M. (2019). Reasoning about reactions in organic chemistry: Starting it in general chemistry. *Journal of Chemical Education, 96*(2), 213–226. https://doi.org/10.1021/acs.jchemed.8b00784
Crandell, O. M., Lockhart, M. A., & Cooper, M. M. (2020). Arrows on the page are not a good gauge: Evidence for the importance of causal mechanistic explanations about nucleophilic substitution in organic chemistry. *Journal of Chemical Education, 97*(2), 313–327. https://doi.org/10.1021/acs.jchemed.9b00815
diSessa, A. A. (2014). A history of conceptual change research. In R. K. Sawyer (Ed.), *The Cambridge handbook of the learning sciences* (2nd ed., pp. 88–108). Cambridge University Press. https://doi.org/10.1017/CBO9781139519526.007
Franovic, C. G.-C., Noyes, K., Stoltzfus, J. R., Schwarz, C. V., Long, T. M., & Cooper, M. M. (2023). Undergraduate chemistry and biology students' use of causal mechanistic reasoning to explain and predict preferential protein–ligand binding activity. *Journal of Chemical Education, 100*(5), 1716–1727. https://doi.org/10.1021/acs.jchemed.2c00737
Glass, A. L., & Kang, M. (2022). Fewer students are benefiting from doing their homework: An eleven-year study. *Educational Psychology*. https://www.tandfonline.com/doi/abs/10.1080/01443410.2020.1802645
Graulich, N., & Caspari, I. (2020). Designing a scaffold for mechanistic reasoning in organic chemistry. *Chemistry Teacher International, 3*(1), 19–30. https://doi.org/10.1515/cti-2020-0001
Graulich, N., & Schween, M. (2018). Concept-oriented task design: Making purposeful case comparisons in organic chemistry. *Journal of Chemical Education, 95*(3), 376–383. https://doi.org/10.1021/acs.jchemed.7b00672
Hammer, D. (2000). Student resources for learning introductory physics. *American Journal of Physics, 68*(S1), S52–S59. https://doi.org/10.1119/1.19520
Hammer, D., Elby, A., Scherr, R. E., & Redish, E. F. (2005). Resources, framing, and transfer. In *Transfer of learning from a modern multidisciplinary perspective* (pp. 89–119). Information Age Publishing.
Hattie, J., & Timperley, H. (2007). The power of feedback. *Review of Educational Research, 77*, 81–112.
Houchlei, S. K., Bloch, R. R., & Cooper, M. M. (2021). Mechanisms, models, and explanations: Analyzing the mechanistic paths students take to reach a product for familiar and unfamiliar organic reactions. *Journal of Chemical Education, 98*(9), 2751–2764. https://doi.org/10.1021/acs.jchemed.1c00099
Houchlei, S. K., Crandell, O. M., & Cooper, M. M. (2023). "What about the students who switched course type?": An investigation of inconsistent course experience. *Journal of Chemical Education, 100*(11), 4212–4223. https://doi.org/10.1021/acs.jchemed.3c00345
Jonsson, A. (2013). Facilitating productive use of feedback in higher education. *Active Learning in Higher Education, 14*(1), 63–76. https://doi.org/10.1177/1469787412467125
Judd Jr., O. H. (2019). *Characterizing student thinking about solutions and the solvation process: The search for mechanistic understanding* [Dissertation]. Michigan State University.
Krist, C., Schwarz, C. V., & Reiser, B. J. (2019). Identifying essential epistemic heuristics for guiding mechanistic reasoning in science learning. *Journal of the Learning Sciences, 28*(2), 160–205. https://doi.org/10.1080/10508406.2018.1510404
Matz, R. L., Fata-Hartley, C. L., Posey, L. A., Laverty, J. T., Underwood, S. M., Carmel, J. H., Herrington, D. G., Stowe, R. L., Caballero, M. D., Ebert-May, D., & Cooper, M. M. (2018). Evaluating the extent of a large-scale transformation in gateway science courses. *Science Advances, 4*(10), eaau0554. https://doi.org/10.1126/sciadv.aau0554
McNeill, K. L., Lizotte, D. J., Krajcik, J., & Marx, R. W. (2006). Supporting students' construction of scientific explanations by fading scaffolds in instructional materials. *Journal of the Learning Sciences, 15*(2), 153–191. https://doi.org/10.1207/s15327809jls1502_1
Messeri, L., & Crockett, M. J. (2024). Artificial intelligence and illusions of understanding in scientific research. *Nature, 627*(8002), 49–58. https://doi.org/10.1038/s41586-024-07146-0

Mislevy, R. J., Almond, R. G., & Lukas, J. F. (2003). *A brief introduction to evidence-centered design* (Research Report RR-03-16). Educational Testing Service.

Moreira, P., Marzabal, A., & Talanquer, V. (2018). Using a mechanistic framework to characterise chemistry students' reasoning in written explanations. *Chemistry Education Research and Practice, 20*(1), 120–131. https://doi.org/10.1039/C8RP00159F

Noyes, K., & Cooper, M. M. (2019). Investigating student understanding of London dispersion forces: A longitudinal study. *Journal of Chemical Education, 96*(9), 1821–1832. https://doi.org/10.1021/acs.jchemed.9b00455

Noyes, K., McKay, R. L., Neumann, M., Haudek, K. C., & Cooper, M. M. (2020). Developing computer resources to automate analysis of students' explanations of London dispersion forces. *Journal of Chemical Education, 97*(11), 3923–3936. https://doi.org/10.1021/acs.jchemed.0c00445

Noyes, K., Carlson, C. G., Stoltzfus, J. R., Schwarz, C. V., Long, T. M., & Cooper, M. M. (2022). A deep look into designing a task and coding scheme through the lens of causal mechanistic reasoning. *Journal of Chemical Education, 99*(2), 874–885. https://doi.org/10.1021/acs.jchemed.1c00959

Pellegrino, J. W., Chudowsky, N., & Glaser, R. (2001). *Knowing what students know: The science and design of educational assessment*. National Academies Press. http://www.nap.edu/catalog/10019

Reiser, B., & Tabak, I. (2014). Scaffolding. In *The Cambridge handbook of the learning sciences* (pp. 44–62). Cambridge University Press.

Russ, R. S., Scherr, R. E., Hammer, D., & Mikeska, J. (2008). Recognizing mechanistic reasoning in student scientific inquiry: A framework for discourse analysis developed from philosophy of science. *Science Education, 92*(3), 499–525. https://doi.org/10.1002/sce.20264

Sadler, D. R. (1989). Formative assessment and the design of instructional systems. *Instructional Science, 18*(2), 119–144. https://doi.org/10.1007/BF00117714

Schwarz, C. V., Cooper, M. M., Long, T. M., Trujillo, C. M., Noyes, K., de Lima, J., Kesh, J., & Stoltzfus, J. (2020). Mechanistic explanations across undergraduate chemistry and biology courses. In M. S. Gresalfi, & I. S. Horn (Eds.), *The proceedings from the fourteenth international conference of the learning sciences (ICLS) 2020* (Vol. 1, pp. 625–628). International Society of the Learning Sciences. https://doi.org/10.22318/icls2020.625

Shepard, L. A. (2000). The role of assessment in a learning culture. *Educational Researcher, 29*(7), 4–14. https://doi.org/10.3102/0013189X029007004

Shepard, L. A. (2005). Linking formative assessment to scaffolding. *Educational Leadership, 63*(3), 66–70.

Uminski, C., Hubbard, J. K., & Couch, B. A. (2023). How administration stakes and settings affect student behavior and performance on a biology concept assessment. *CBE—Life Sciences Education, 22*(2), ar27. https://doi.org/10.1187/cbe.22-09-0181

Underwood, S. M., Kararo, A. T., Posey, L. A., Pollock, A. M., Herrington, D. G., Stowe, R. L., Carmel, J. H., Klymkowsky, M. W., & Cooper, M. M. (2023). Components critical to successful adoption and adaptation of CLUE, a transformed general chemistry curriculum. *Journal of Chemical Education*. https://doi.org/10.1021/acs.jchemed.3c00190

van de Pol, J., Volman, M., & Beishuizen, J. (2010). Scaffolding in teacher–student interaction: A decade of research. *Educational Psychology Review, 22*(3), 271–296. https://doi.org/10.1007/s10648-010-9127-6

van Mil, M. H. W., Boerwinkel, D. J., & Waarlo, A. J. (2013). Modelling molecular mechanisms: A framework of scientific reasoning to construct molecular-level explanations for cellular behaviour. *Science & Education, 22*(1), 93–118. https://doi.org/10.1007/s11191-011-9379-7

Winstone, N. E., Nash, R. A., Rowntree, J., & Parker, M. (2017). 'It'd be useful, but I wouldn't use it': Barriers to university students' feedback seeking and recipience. *Studies in Higher Education, 42*(11), 2026–2041. https://doi.org/10.1080/03075079.2015.1130032

Wood, D., Bruner, J. S., & Ross, G. (1976). The role of tutoring in problem solving. *Journal of Child Psychology and Psychiatry, 17*(2), 89–100. https://doi.org/10.1111/j.1469-7610.1976.tb00381.x

Chapter 7
Explicating Black Boxes in the Biology Classroom for Promoting Mechanistic Reasoning

Gur Arie Livni Alcasid and Michal Haskel-Ittah

7.1 Introduction

7.1.1 Explanatory Black Boxes in Mechanistic Explanations

Mechanistic explanations reveal the rich causal structure underlying phenomenon by detailing the way entities and their activities produce them (Craver & Darden, 2013, p. 65). Alongside this richness, mechanistic explanations often (or always, as we are about to argue) include black boxes, knowledge gaps in the description of the mechanism (Haskel-Ittah, 2023). As mechanistic explanations enable us to explain causal relations as a link between cause and effect, such black boxes are the missing parts in a complete causal chain of events (Machamer et al., 2000).

In our discussion of black boxes in mechanistic explanations, we focus on absent details that serve to understand a mechanism. Not every missing detail qualifies as a black box; for instance, omitting the number of cells in a tumor or the scientific name of a risk factor does not. However, withholding key information—like *how* a chemical causes DNA mutations—does represent black boxing.

While black boxes might be considered an impede to the integrity of a mechanistic explanation, we argue that these black boxes are an inevitable and productive part of every mechanistic explanation. As an example, let us examine an explanation of the body's reaction to sugar consumption. This phenomenon includes increased blood glucose levels (input) and subsequent regulation, mediated by glucose entrance into the cells (output). Here, the whole mechanism is hidden in a black box. Unpacking this phenomenon black box may involve explaining that high

G. A. Livni Alcasid · M. Haskel-Ittah (✉)
Department of Science Teaching, Weizmann Institute of Science, Rehovot, Israel
e-mail: Michal.Haskel@weizmann.ac.il

N. Graulich et al. (eds.), *Exploring Mechanistic Reasoning in Science Education*, Contributions from Science Education Research 17,
https://doi.org/10.1007/978-3-032-10246-1_7

glucose levels lead to insulin release from the pancreas, causing a decrease in blood glucose levels. While this explanation unpacks some entities and activities, other processes along the mechanism are kept as explanatory black boxes, such as how the pancreas senses and secretes insulin and how insulin reduces blood glucose levels. Unpacking them will reveal even more black boxes. For example, explaining that insulin activates membrane receptors, which signal for recruitment of glucose transporters, reveals additional black boxes, such as the insulin-receptors biochemical interaction or the signaling process.

Considering black boxes as an informational unit in mechanistic knowledge, science often sets their unpacking as a goal. Scientific knowledge is accumulated by unpacking black boxes, regardless of how this information is used in explanations (Machamer et al., 2000). However, the description of a mechanism and the construction of a mechanistic explanation has its own epistemic value in science, as it allows for the evaluation of the proposed mechanism (Bechtel, 2007). Paradoxically, once unpacked, entities can be further decomposed into subparts, their activities into subactivities, transforming mechanisms into more fine-grained sub-mechanisms. Therefore, unpacked black boxes will always produce more emergent black boxes. Eventually, endless unpacking could lead to a cascade of reduction that results in an overly complex and ultimately unproductive explanation (Băetu, 2015; Nathan, 2021).

It should now be evident that the construction of mechanistic explanations will always lead to the emergence of black boxes. Haskel-Ittah (2023) termed these emerging black boxes as *explanatory black boxes*, describing them as arising from two distinct origins: the nature of science and the nature of mind. In science, explanatory black boxes represent gaps in mechanistic knowledge, often stemming from unknown details awaiting further research, such as the protein-production-mechanism black box studied throughout the twentieth century. In the realm of mind, limited human cognitive capacity often requires incomplete explanations (Keil, 2019), which contain multiple black boxes. This need for conciseness not only requires omitting details but also demands a degree of abstraction to avoid overwhelming complexity.

In our earlier example, let us assume an aim to explain how a lack of insulin leads to high blood glucose levels. Discussing the pancreas's insulin secretion mechanisms doesn't clarify this. Similarly, examining glucose's atomic structure and its interactions with membrane transporters would complicate our understanding of glucose regulation.

Explanatory black boxes are not only a unit hiding details but also a unit supporting focus and clarity. Given that mechanisms are not just processes that produce a phenomenon but also structured explanations of those processes (Illari & Williamson, 2012), constructing a mechanism inevitably involves the use of black boxes, preventing mechanistic explanations from becoming overly detailed and thus incomprehensible (Băetu, 2015; Nathan, 2021).

To sum up, explanatory black boxes are mechanistic knowledge gaps residing in mechanistic explanations. They are termed explanatory because (a) they are the result of an attempt to describe the contents of a black box, (b) they are an inevitable and integral part of mechanistic explanations, (c) they serve a productive role in the construction of mechanistic explanations and (d) their unpacking results in the revelation of additional explanatory black boxes.

7.1.2 Explanatory Black Boxes in Science Education Research

In recent years, science education research has provided important insights into the affordances and limitations of mechanistic reasoning in the science classroom. By focusing on the nature of unpacking in various science domains, researchers have advanced our understanding of students' ability to comprehend and explain natural phenomena (e.g., de Andrade et al., 2022; e.g., Dickes et al., 2016; Duncan & Tseng, 2011; Kapon, 2017; Krist et al., 2018; Talanquer, 2010; Weinberg, 2018).

Bachtiar et al. (2022), reviewed 60 papers on mechanistic reasoning from 2006 to 2021, demonstrating unpacking as a key theme in them. Most papers view mechanistic reasoning as a cognitive process identifying essential components like entities and activities. Many papers note that students find unpacking challenging, leading to interventions aimed at helping them learn this skill. Thus, mechanistic reasoning is almost conjoined in these papers with unpacking. However, considering the inevitability of explanatory black boxes in every mechanistic explanation, we recognize a crucial element seldom discussed.

We acknowledge and understand the importance of assisting students in unpacking mechanistic information. However, we suggest that while unpacking may be a goal in teaching mechanisms in the classroom, expert unpacking is not the indiscriminate unveiling of entities. Cooper & Noyes (2025) address this challenge in Chapter 6 of this book, explaining how in chemistry education students need guidance in choosing the correct organizational level for their explanations rather than defaulting to the lowest level. Similarly, our own experience with teaching biology has shown that complex mechanisms often operate across multiple organizational levels and include numerous unobservable entities and activities (Haskel-Ittah et al., 2020; Haskel-Ittah & Yarden, 2018). This complexity limits students' ability to reason about such mechanisms, but it also highlights the need to evaluate which mechanistic details enhance comprehensibility and which may be better left black-boxed.

Whether black boxing decisions in the science classroom should be identical or even alike black boxing decisions made by scientists is a significant question. Both science and the science classroom position the formulation of explanations as an important practice, and while some researchers assert that they are not the same, most agree that there are at least some similarities between them (Alameh et al., 2023). The context of black boxes adds complexity to this question, as the productivity of unpacking is highly dependent on explanation goals and audience.

Science and the science classroom sometimes adopt different goals and practices to achieve them (Alameh et al., 2023; Berland et al., 2016). Specifically in the context of black boxes, we observed that teachers' reasons for unpacking and black boxing mechanistic details sometimes concur with scientific considerations but are often affected by other contradictory factors (Livni Alcasid & Haskel-Ittah, 2024).

For example, teachers adjust the level of unpacking in their classroom according to a specific course's goal, alignment with learning progressions, and how they perceive their students' motivation and cognitive level.

Contrarily, this chapter argues for the value of finding parallels between mechanistic reasoning in scientific and educational settings, particularly regarding the identification and role of explanatory black boxes. In science, identifying black boxes propels inquiry, prompting further research to clarify these gaps (Craver & Darden, 2013; Nathan, 2021). Similarly, teachers can use black boxes in the classroom to spark student curiosity and link prior knowledge to new concepts, fostering an inquiry-based approach (Livni Alcasid & Haskel-Ittah, 2024; Olsher & Dreyfus, 1999).

In science, constructing mechanistic explanations involves skillfully applying black boxes to create clear, productive explanations that balance completeness and simplicity. Conversely, science also holds instances where scientists dismissed the importance of recognized black boxes, leading them to oversee important aspects of scientific inquiry, hindering mechanistic reasoning and explanation epistemology (Băetu, 2015; Nathan, 2021). This means that the selection and placement of explanatory black boxes in science entails a complex and skillful technique.

In the classroom, it remains an open question whether the discussion of black boxes can produce a more nuanced and productive reasoning. Therefore, this study aimed to investigate whether incorporating black-boxing practices could support students' mechanistic reasoning and facilitate a closer alignment with scientific inquiry practices.

While most research on mechanistic reasoning focuses on unpacking mechanisms, some recent studies have shifted attention to black boxes. Two recent studies targeted the teaching of a black box on the borderline between biology and chemistry. This black box is the implied black box in the wording 'ATP *provides* energy' (Franovic et al., 2023; Talanquer & Szozda, 2024). 'Providing' is a filler term masking a complex biochemical mechanism. Such masking assists students by simplifying the process but may lead to misconceptions (Talanquer & Szozda, 2024). By interviewing chemistry, biology, and biochemistry instructors, Franovic et al. (2023) found that instructors relied primarily on two mechanisms to explain this filler term: one focusing on ATP hydrolysis and the other on phosphorylation. The latter was suggested as more productive for students and less demanding for teachers. In addition, the authors suggested that this black box should be explicated to emphasize its existence in this reaction before its unpacking.

Other studies in science education also supported the explication of explanatory gaps as a tool to focus students on unpacking (Gericke & Mc Ewen, 2023; Jahanifar & Masnavi, 2023; Olsher & Dreyfus, 1999; Tang & Hammer, 2024). These studies suggest that assisting students in recognizing explanatory black boxes can support their mechanistic reasoning. Olsher & Dreyfus (1999) demonstrated the effectiveness of an iterative strategy, where teachers directed students to ask unpacking questions. However, none of these studies address the explicit use of the black box term in the classroom, i.e., making students aware of black boxes as an abstract, yet inevitable, construct in mechanistic explanations.

In a recent study, we investigated the complexity of teachers' decision-making processes regarding the unpacking and black boxing of mechanistic details in classroom explanations. Discussing the black box term explicitly with teachers revealed a range of teacher considerations, including students' cognitive levels, curriculum constraints, and efforts to convey the nature of science. Furthermore, we discovered that teachers perceived the explication of black boxes as a method to leverage black boxing to attain pedagogical goals, such as creating curiosity, linking topics across the curriculum, and assisting students in understanding mechanistic complexity while avoiding unnecessary unpacking (Livni Alcasid & Haskel-Ittah, 2024).

This led us to reconsider another potential outcome of explaining black boxes to students. Since unpacking or black boxing involves decision-making processes, we hypothesized that explicating such processes for students may help them develop mechanistic explanations.

7.1.3 The Potential of Discussing Black Boxes with Students

We have previously witnessed that students can appreciate the productiveness of black box explanations when explaining phenomena (Livni Alcasid & Haskel-Ittah, 2022). Emerging evidence suggests that supplementing students' experiences with reflective discussions on epistemological issues is paramount for nurturing a profound comprehension of the nature of science (Abd-El-Khalick, 2013; Berland et al., 2016; Schwartz et al., 2004). Considering, it could be hypothesized that explicitly discussing the existence of black boxes could improve students' understanding of epistemic issues involved in constructing mechanistic explanations. This means recognizing the contribution of unpacking mechanistic elements to the explanation's productivity, but also more intricate issues, such as the appropriateness of specific black boxes in certain explanations and the value of explanations even as they contain black boxes.

In contrast, the abstract nature of the 'black-box' term raises questions about students' ability to understand and apply it fully. Moreover, science students are rarely in a position where they need to decide which information is relevant to their explanation construction. Studies suggest that such decisions are typically predefined by the curriculum or guided by the teacher (Berland & Hammer, 2012; Livni Alcasid & Haskel-Ittah, 2024; Miller et al., 2018; Tang et al., 2020). Therefore, while scientists are expected to engage in reflective decision-making about what to unpack, it remains unclear if today's science classrooms inherently facilitate such discussions or what aspects of classrooms could support them.

Our primary goal in this study was to determine the feasibility of a black-box discourse in the science classroom. Following this, we were also interested in examining whether such a discourse can promote meaningful scientific discussion within the classroom.

To achieve these goals, we realized that investigating the applicability of a black box discourse in the classroom requires examining both students' proficiency in understanding the term and using it to discuss unpacking and black-boxing options in constructing mechanistic explanations.

Therefore, to investigate these issues, this study posed the following research questions (RQ):

1. How do students interpret the ‘black-box’ term in the context of constructing mechanistic explanations?
2. How do students discuss the ‘black-box’ term in the context of constructing mechanistic explanations?

7.2 Study Context

7.2.1 Participants and Intervention

To examine students’ understanding and application of the ‘black-box’ term, we developed a special unit focused on explicating black boxes within the context of cancer development. The unit was designed to teach 10th-grade biotechnology students about the mechanisms underlying cancer onset while iteratively highlighting black boxes. In alignment with our understanding of the black-box term, the course directed students to identify mechanistic gaps in their understanding of cancer biology, referred to explicitly as “black boxes.” This design aimed to encourage deeper engagement with the underlying mechanisms, helping students move beyond surface-level correlations to a more mechanistic understanding of cancer-related phenomena.

To exemplify our use of black box explication, Fig. 7.1 illustrates the third chapter of the unit, focusing on the way black boxes were integrated into the course for

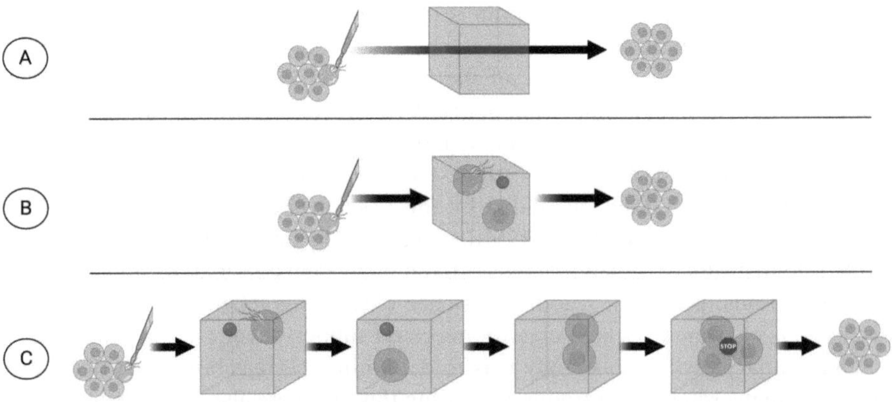

Fig. 7.1 An illustration of the wound healing mechanism from the unit’s third chapter. As students advanced through the course, knowledge was represented by black boxes that were unpacked iteratively. The chapter started with the relationship between tissue healing and cell proliferation (**a**), unpacked the role of growth factor secretion (**b**), leading to more explanatory black boxes (**c**)

students. As students learned about cell proliferation in wound healing, the "black box" term was used to highlight the mechanistic knowledge gap in this input-output relationship (a in Fig. 7.1).

The chapter unpacked specific activities, such as damaged cells secreting growth factors that lead to cell proliferation (b in Fig. 7.1). This unpacking was followed by an explication of several emergent black boxes related to growth factor secretion, signal transduction, cell proliferation, and regulatory mechanisms that stop proliferation (c in Fig. 7.1).

Students were then asked to discuss which black box should be unpacked to explain why cancer cells proliferate in the absence of growth factors. Following, the chapter proceeded to unpack the signal transduction pathway involving the growth factors and receptors.

Throughout the unit, students were repeatedly asked to write how their newly acquired knowledge assists them in explaining the original phenomenon and suggest further questions that can be asked considering their newly acquired understanding.

A year before this study, the teacher delivering the course underwent a 30-hour professional development program to support the course's facilitation. This included activities supporting the teaching of mechanisms, including familiarity with the 'explanatory black boxes' term.

The course spanned approximately 14 hours over 7 weeks, with students participating in weekly 2-hour sessions. The course was delivered in the teachers' classroom. The teacher utilized school computers to provide students access to online course materials via PeTeL, a Moodle-based learning management system developed by the Department of Science Teaching at the Weizmann Institute of Science. This blended approach allowed for a combination of in-person instruction and online interactive learning.

At the end of the unit, the teacher assigned students a summative task addressing her pedagogical needs; she instructed students to create an informational poster on a disease of their choice. For this task, students worked in groups of 1–3, researching and integrating online information. The teacher's guidance focused on poster structure and source reliability.

Students chose a variety of diseases for their projects. Six groups selected different types of cancer: colon cancer, glioblastoma, leukemia, pancreatic cancer, pineoblastoma, and squamous cell carcinoma. Five groups chose other non-cancer diseases: diabetes, flu, Parkinson's disease, rabies, and smallpox.

7.2.2 Data Collection

To address the research questions, we used two learning artifacts created by students to assess evidence for their learning, i.e., posters and reflection sheets. The first artifact was the informational posters students created as the summative assignment. The second artifact was a worksheet given to the students at the beginning of their work on the posters. These worksheets instructed students to specify at least one mechanism they unpacked and one mechanism they black boxed in their poster and elaborate on their considerations to do so. Students filled out these worksheets at the end of the activity, using them to reflect on their work processes.

As noted, students chose a variety of diseases for their projects, creating a challenge for data analysis. To account for different phenomena described at varying organizational levels, we used a coding scheme that was both flexible enough to address each case and rigorous enough to enable meaningful comparisons. The following sections present the data analysis and results for each research question.

7.3 RQ1 How Do Students Interpret the Black-Box Term in the Context of Mechanistic Explanation Construction?

7.3.1 Data Analysis

In our first research question, we aimed to determine whether students' use of the terms "black-boxing" and "unpacking" aligned with their traditional scientific meanings. In their worksheets, students indicated which sections of their informational posters contained unpacked or black-boxed information. To address our research question, we wanted to examine whether students' classification of explanation types—as reflected in their worksheets—matched the conventional meaning of the "black-boxing" and "unpacking" terms.

For each mechanism reported in the worksheet, we located the relevant section in the corresponding poster to determine if it was unpacked or black boxed by the students. To code the data, we adapted the coding scheme of Russ et al. (2008), which classifies the actions of *describing the target phenomenon* and *identifying setup conditions* as mechanistic reasoning procedures differing from unpacking. These actions were coded as *black boxing*. Additionally, Russ et al.'s framework describes three unpacking actions, including *identifying entities*, *identifying activities,* and *chaining: backward and forward* as inherently hierarchical. We included these as three separate hierarchical levels of *unpacking* to denote students' attempts at unpacking with varying degrees of sophistication.

To elucidate the three hierarchical levels of unpacking, the lowest level denoted mechanistic explanations including only entities, the second level mechanistic explanations including entities and their activities, and the third level explanations including chaining—the linking of activities to broader aggregated phenomena—a critical step in ensuring causal coherence in mechanistic explanations, as discussed by (Krist, 2020).

Additionally, a bottom-up analysis uncovered five cases referring to the unpacking and black boxing of non-mechanistic details. For example, students wrote they "black-boxed" the cost of a treatment or "unpacked" the procedure of colonoscopy. We identified these as students' attention to adding or subtracting details. However,

Table 7.1 Coding scheme for characterizing explanation type (non-mechanistic, black box and mechanistic) in students' posters

Code category	Code and code description	Example
Black boxing	*Black boxing*: Students included only phenomenon description, input and output of phenomenon or a description of sequenced events, without unpacking the causality between them	The immune response includes the creation of memory cells; 48 hours after vaccine treatment, side effects, such as headache or eye swelling may appear
Unpacking	*Unpacking level 1:* Students described an entity in an organizational level below the target phenomenon	Cigarettes contain toxic chemicals; This treatment is based on human antibodies
	Unpacking level 2: Students described an entity in a organizational level below the target phenomenon, and its activity	At this stage, the body produces the protein interferon which warns the immune system of a virus
	Unpacking level 3: Students linked underlying activities to the overarching phenomenon	Lack of dopamine causes neurons to transmit random stimuli, which leads to uncontrolled movements
Non-mechanistic	*Non-mechanistic*: Students discussed a non-mechanistic detail	The researchers used a special animal model; Cost of study

these did not include any mechanistic details in the context of biological mechanisms. Accordingly, they were coded as *non-mechanistic*. Table 7.1 elaborates our adapted coding scheme.

After developing the coding scheme, both authors discussed it to enhance its validity. After reviewing 10% of the data together, each author independently coded a random sample comprising 20% of the data. Krippendorff's Alpha was calculated to assess inter-rater reliability, resulting in a coefficient of 0.83. Since this exceeds the 0.8 threshold for satisfactory reliability (Krippendorff, 2019), the first author proceeded to code the remaining data.

To clarify our methodology, we use two examples, one of a group discussing rabies in their poster and the other of a group discussing leukemia. Figure 7.2 illustrates the first example. In this example, we examined the group's worksheets and discovered that they classified their explanation of how the rabies virus travels through the body as an unpacked mechanism. We then located the relevant section in the group's poster, reviewed it, and coded it as a level 3 unpacked mechanistic explanation. It can be observed that this case yielded an alignment between students' classification and our coding.

In a more complex example, the group discussing leukemia classified an explanation of the disease risk factors appearing in their poster as unpacked explanations. Upon examining the section of the poster about risk factors, it was found to contain seven different explanations, each corresponding to a specific risk factor. We coded each explanation individually, identifying which were black boxes, which included unpacking (levels 1, 2, or 3), and which contained non-mechanistic details. Out of the seven explanations, we classified three as black boxes, one as a level 1 unpacking, two as level 2 unpacking, and one as a level 3 unpacking. This approach enabled

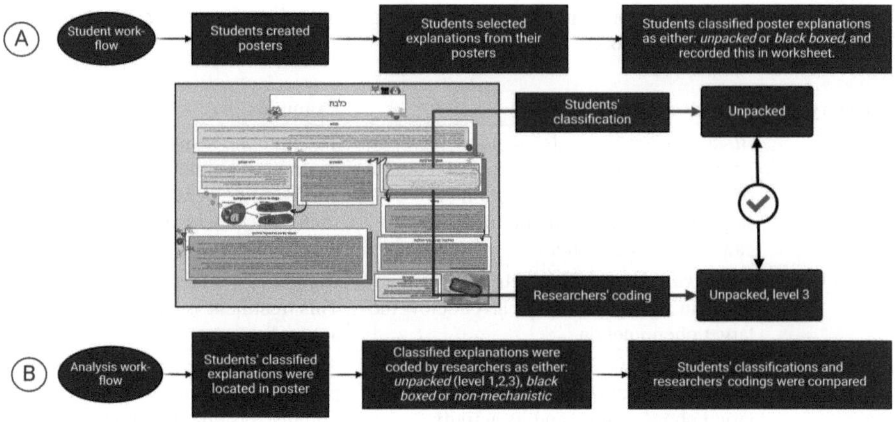

Fig. 7.2 Data analysis process for RQ1. (**a**) Students recorded their classifications of poster explanations in worksheets. (**b**) Following, the researchers examined the worksheets, located the original explanations in the poster and coded them. Once the explanations in the poster were coded, the codes were compared with the students' classifications

us to uncover varying degrees of alignment between the students' classifications and our coding.

By cross-referencing the classifications recorded in students' worksheets with the posters, we compiled a detailed list of 85 items. Each item represented a mechanism discussed, the students' classification (i.e., unpacking or black boxing), and the code assigned to the corresponding explanation by us.

7.3.2 Results for RQ1

Using the Russ et al. (2008) framework, we compared students' classifications of explanations (i.e., unpacking or black boxing) with our coding of these explanations. Figure 7.3 shows the distribution of our coding in relation to the two separate classifications used by the students. Data indicate that 72.73% of explanations classified as unpacked by the students contained some degree of unpacking. Similarly, 76.67% of explanations classified as black boxed were aligned with our coding of black boxing. Misalignment was observed mainly when students classified explanations as unpacking, while we identified these as black boxed (25.45%), and when students classified explanations as black boxed, while we identified these as non-mechanistic (16.66%). We also note that within correctly aligned unpacking classifications, only 50% were identified as high-level unpacking (i.e., level 3). In other words, even when students used the 'unpacking' term correctly, an observable proportion of their explanations included sub-optimal unpacking.

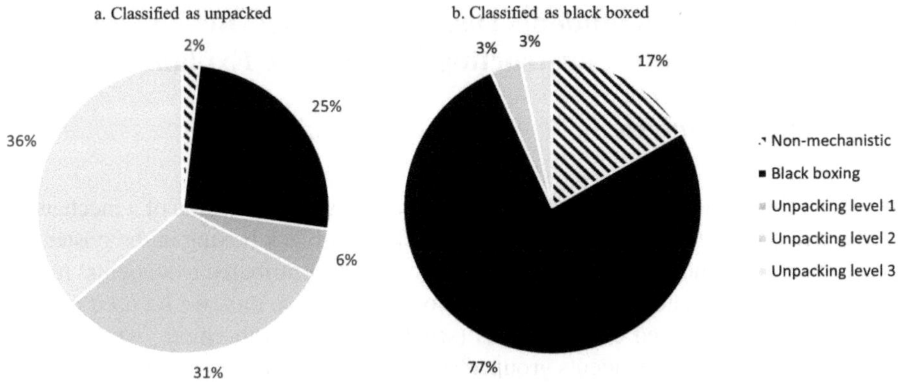

Fig. 7.3 Distribution of coded explanations in relation to students' classification of explanations (unpacking or black boxing) (n = 85). Data demonstrates a large proportion (73%) of unpacked explanations in explanations students classified as unpacked (**a**) and a large proportion (77%) of black boxed explanations in explanations students classified as black boxed (**b**)

Table 7.2 Alignment of students' classification of unpacking and black boxing explanations with codes given by authors

		Aligned claim	Misaligned claim	Total
Claimed explanation	Black boxing	23	7	30
	Unpacking	40	15	55
	Total	63	22	85

To examine the significance of the results, we first categorized students' classification of explanations (i.e., black box or unpacking) as aligned or misaligned with our coded explanations (Table 7.2). Next, as the proportions between aligned and misaligned classifications were similar for both black boxing and unpacking (76.67% and 72.73%), we performed a Chi-Square Goodness-of-Fit test, considering the proportion of all aligned classifications (63 of 85). This test was performed to evaluate whether rates of aligned classifications were significantly different from chance (50%). The observed frequencies of correctly identified explanations were compared to the expected frequencies under random assignment. The results of the Goodness-of-Fit test showed a significant deviation from random assignment ($\chi^2 = 14.412$, $p = 0.00015$). These results suggest that students were significantly more accurate than chance in classifying both black boxing and unpacking explanations.

Taking the current data into account, we concluded that while some students interpreted the "black-boxing" and "unpacking" terms in a less conventional manner, most students interpreted them as intended. However, once they tried to unpack the mechanisms they intended to unpack, only half included high-level unpacking.

As we proceed to our following research question, we will illustrate two distinct ways by which students used the black box term to discuss their construction of mechanistic explanations and choices of black boxing (RQ2).

7.4 RQ2 How Do Students Discuss the 'Black-Box' Term in the Context of Constructing Mechanistic Explanations?

7.4.1 Data Analysis

As mentioned, students' worksheets included students' identification of a mechanism and a written justification supporting its unpacking or black boxing in the poster. We aimed to uncover how the use of the 'black-box' term contributed to students' review of their unpacking and black boxing decisions. To achieve this, we focused only on cases discussing aligned classifications (see Table 7.2). Furthermore, while posters contained 85 items, the students grouped some items (such as "risk factors") into one in the worksheet. These led to a reduction in our sample from 85 to 40 items.

To systematically analyze the data, each time students discussed a mechanism, we treated it as a single unit of analysis, even if it contained several independent considerations. By evaluating the entire paragraph or explanation as a unit, we ensured that we captured the complete reasoning and context behind the students' choices while preventing fragmentation of students' written justifications.

Following, we adopted a bottom-up approach, allowing themes to emerge from the data. This ensured a reflection of students' thinking, capturing the complexity and context of their justifications. Given the distinctiveness of this intervention, we wanted to ensure our analysis captured any unexpected patterns or themes that may not have been anticipated during our initial research design.

Our bottom-up analysis revealed two perspectives in students' discussions. The first was a *single black box perspective* in which students introduced considerations supporting the unpacking or black boxing of one mechanism. The second was the *interconnected black boxes perspective*, in which students introduced considerations of comparing two or more black boxes. We exemplify and discuss these two categories in the results section.

Due to the small sample size used for this research question, the first author coded all the data after developing the coding scheme. The second author was then given eight student statements that had been coded as a single black box perspective, as well as eight statements coded as interconnected black box perspectives, without being informed of which was which. After the second author completed the coding, the two sets were compared, revealing perfect agreement between the two authors.

Following the coding, we proceeded to quantify the statements by analyzing their distribution in relation to the prompts issued to the students (e.g., justify unpacking or justify black boxing).

7.4.2 Results for RQ2

Our bottom-up analysis revealed two distinct ways in which students presented their unpacking and black boxing considerations: a. *single black box perspective* and b. *interconnected black boxes perspective.* Generally speaking, the single black box perspective involved students' discussion of one explanatory black box. They provided justifications regarding unpacking or black boxing that particular explanatory black box. In the interconnected black boxes perspective, students still focused on one explanatory black box. However, they compared it to other explanatory black boxes while doing so. To elucidate these two distinct perspectives, we provide several examples.

The single black box perspective involved introducing one or several considerations providing direct support to students' decisions. For example, the group discussing pineoblastoma (a type of brain tumor) justified its decision to black box the disease onset mechanism by explaining that this mechanism is unknown to science:

> The cause and reason for pineoblastoma onset are unknown… This box stayed closed as we were left out of options.

In other cases, students used several independent considerations to justify unpacking or black boxing. For example, a student discussing the flu mentioned three different considerations (teacher expectations, poster length, and poster focus) to support his decision to black box the side effects of the flu vaccine, by writing:

> It's complicated, and I don't think the teacher wanted me to detail it. It's definitely going to be long, and I want to stay focused on the flu.

In one unique item, we identified a student considering conflicting considerations, demonstrating a more complex approach to her decision. In this case, the student presented considerations for both unpacking and black boxing the squamous-cell-carcinoma onset mechanism:

> In a science project aimed at a reader with an understanding of biology, it is important to mention not just the risk factors, but also the mechanism which creates cancer… to the general public, I wouldn't specify a biological mechanism as it would be unnecessary information that detracts from the message.

Through our analysis, we uncovered similar statements exemplified by the cases above, focusing on one mechanism, while directly supporting its unpacking or black boxing. We termed this perspective as a single black box perspective.

Contrary to the single black box perspective, we identified an alternative perspective that involved the introduction of several considerations to create a comparative analysis between two or more black boxes. For example, in the group discussing pancreatic cancer, the students justified black boxing the erlotinib EGFR inhibition mechanism, by comparing its productivity to that of other parts of the disease mechanism:

> We left this mechanism black boxed because we chose to detail the mechanisms that are more important for understanding the drug. Therefore, we chose not to explain how the binding of the active substance to tyrosine kinase impairs the activity of the receptor protein, since it was possible to understand the drug's mechanism without this explanation.

In this example, the group discusses their decision to black box one mechanism. However, to do so, they first discuss their decision to unpack other mechanisms deemed as more important for understanding. Following, the group continues to compare the productivity of both, stating that the discussed mechanism is less productive in supporting their goals.

Table 7.3 Distribution of local perspective and global perspective when justifying black boxing and unpacking choices

	Single black box	Interconnected black boxes
Unpacking	19	0
Black boxing	13	8

Another unique case was revealed in a worksheet discussing leukemia. In this worksheet, the participating students wrote that "*the poster includes information about a lot of different blood cells but no information on each of them separately because there are so many types and most of them are irrelevant.*" In this case, students described a compromise in which the abundance of similar black boxes can't permit their ubiquitous unpacking, thus using comparative analysis to create an 'all or none' rule for unpacking.

All cases in this category involved a comparison *between* black boxes, in which students discussed the productiveness or interest provided by the unpacking of several black boxes in relation to each other. We termed this perspective an *interconnected black boxes* perspective.

Analyzing the distribution of perspectives, we found that regardless of prompt, most items contained the single black box perspective. However, on eight occasions, and only when discussing black boxing, students used the interconnected black boxes perspective. Table 7.3 contains the distribution of both perspectives in relation to the prompt.

Fisher's Exact Test was performed to determine if there was a nonrandom association between the prompt (justify unpacking or justify black boxing) and the perspective used (single black box or interconnected black boxes). The test yielded a p-value of 0.0036, indicating a significant association between the explanation type and the perspective students used to justify it. More specifically, data suggest that justifying black boxing facilitated the use of the interconnected black boxes perspective in a way that unpacking did not.

7.5 Discussion

The study detailed in this chapter investigated one facet of black box explication: the inevitable existence of black boxes and considerations for their existence. Given that reflective discussions on epistemological issues are vital for a deep understanding of the nature of science (Abd-El-Khalick, 2013; Berland et al., 2016; Schwartz et al., 2004), this study aimed to determine how highlighting the existence of black boxes to students can support their discussions of black boxing decisions and enhance their understanding of mechanistic explanations.

In this study, we asked students to write their considerations for unpacking and black boxing mechanistic details, following their creation of informational posters. Subsequently, we incorporated the framework of Russ et al. (2008) to determine whether students could classify and identify unpacking and black boxing explanations in their posters. This analysis was followed by a bottom-up approach to examine how students discussed their unpacking and black boxing considerations.

Our data analysis revealed that in most cases, students correctly understood and used the terms "black-boxing" and "unpacking." We have previously demonstrated that students can appreciate the productiveness of black box explanations due to their use in producing predictions (Livni Alcasid & Haskel-Ittah, 2022). However, we couldn't ascertain from previous data whether students can recognize such explanations as black box explanations. Our current study allows us to conclude that high-school students can explicitly recognize explanatory black boxes. Furthermore, they can distinguish between explanatory black boxes and unpacked sections in mechanistic explanations.

Contrarily, this study shows that even when students aim to unpack mechanistic information, they sometimes struggle with the unpacking process. This aligns with previous findings (Bachtiar et al., 2022), confirming what had been suggested *but not clearly demonstrated before-* even students motivated to construct mechanistic explanations may still encounter difficulties in comprehensively and coherently unpacking all the relevant details. Additionally, the data highlights the distinction between explanatory black boxes resulting from students' incomplete unpacking and those deliberately placed by students after careful evaluation.

Alongside these conclusions, we also note that some students' uses of the "unpacking" and "black-boxing" terms were not aligned with conventional meanings. We particularly note our observation of a less conventional use of the term 'black-box,' especially when students referred to non-mechanistic details. A possible explanation may be students' misunderstanding of the 'black-box' term, as they generalized it to mean 'any missing information'. Another possible explanation may be students' inability to distinguish between mechanistic and non-mechanistic details. Although such a task is often regarded as simple (Lockhart et al., 2019), it could be that this task can become challenging in the context of complex biological phenomena or classroom culture.

A relevant question arising from this study is the role of students' prior knowledge and understanding in their ability to identify and use black boxes. In this research, students created explanations for diseases—some directly aligned with the preceding intervention on cancer, while others focused on non-cancer diseases. Although data on students' prior knowledge of these diseases was not collected, their explanations often required integrating new information with familiar mechanisms.

Previous studies highlight that recognizing complex mechanisms often depends on domain-specific knowledge (Haskel-Ittah et al., 2020; Haskel-Ittah & Yarden, 2018). Conversely, research suggests that students can develop mechanistic reasoning skills before constructing explanations that align with scientific norms (Fuhrmann et al., 2024) and can evaluate explanation adequacy even when struggling to generate adequate explanations themselves (Alameh et al., 2023).

The findings of this study suggest that when students acquire a basic understanding of the organizational level discussed, they can recognize explanatory black boxes and evaluate their productivity, even when engaging with new information. These insights underscore the importance of instructional strategies that scaffold domain-specific knowledge while fostering an understanding of explanation structures.

Analyzing the appropriate cases in which students discussed their unpacking and black boxing decisions revealed two distinct perspectives. The first was the single black box perspective, in which students supported their decisions by detailing one or several considerations. The second was the interconnected black boxes perspective, in which students compared the value of unpacking two or more black boxes. When analyzing the distribution of these perspectives in response to the prompts (justifying unpacking or justifying black boxing), it was found that while the single black box perspective was more common for both prompts, the interconnected black boxes perspective appeared only in response to the black boxing prompt.

The interconnected perspective highlights a unique approach to students' evaluation of black boxing decisions. By using a comparative method, students can explain how one explanatory black box supports the unpacking of another. This perspective reveals students' ability to discuss both the inevitability and productivity of black boxes. Inevitability is shown through their use of relative and tentative criteria for unpacking, rather than absolute ones. Productivity is evident in their recognition that black boxing can help in unpacking other boxes.

Conversely, while the single black box perspective can also address these aspects, it doesn't inherently do so. For instance, stating, "*It's definitely going to be long,*" lacks recognition of productiveness. However, reframing this can lead to a richer discussion, such as: "It's definitely going to be long, but I preferred unpacking mechanism X." This emphasizes the importance of recognizing the relative roles of black boxes in constructing coherent explanations.

Previous studies highlight the importance of explicit discussions on epistemological issues in the science classroom (Abd-El-Khalick, 2013; Berland et al., 2016; Schwartz et al., 2004). This study supports these observations. Black box explication and discussion promoted a more complex conversation among students in this study. Specifically, it allowed students to compare the productivity of various mechanistic details to the overall quality of the explanation, thus entailing a discussion of unpacking priorities.

We are limited in determining how exactly the focus on black boxes (as opposed to a focus on unpacking) promotes or supports the interconnected black box perspectives. One possible explanation is that discussing 'what goes in' an explanation narrows students' focus to the merits of unpacking. In contrast, discussing 'what goes out' challenges students to justify not only why specific details are excluded but also how these exclusions enhance the inclusion of others. This explanation seems particularly plausible when students view unpacking as a key goal and regard completely black-boxed descriptions as less explanatory.

Summing up all the above, our study contributes to current mechanistic reasoning frameworks in several ways. First, we wish to emphasize that a significant proportion of student statements included the correct recognition of black boxes within mechanistic explanations, followed by a justification of their existence. This suggests that students sometimes view explanatory black boxes as productive. Such productivity is supported by science philosophy literature (Băetu, 2015; Nathan, 2021) and psychology literature (Keil, 2019).

Mechanistic explanations containing black boxes can be considered less sophisticated (Russ et al., 2008) or a precursor to mechanistic explanations (Krist, 2020). However, we submit that judging students' ability to use mechanistic reasoning cannot be removed from their explanation goals and considerations. This is irrevocably related to using explanatory black boxes in mechanistic explanations.

Our second point is that mechanistic reasoning, as exemplified in this study, includes not only the ability to unpack and use entities. Our study included evidence of students' ability to discuss, scale, compare, and prioritize the unpacking of several black boxes related to each other. Our data support the idea that understanding mechanistic reasoning requires fostering students' ability to navigate the inclusion or exclusion of details within their explanations, contributing to a more nuanced view of their mechanistic reasoning.

As we move forward, we must view black boxes not as mere gaps but as strategic elements that can enhance the clarity and productivity of classroom scientific explanations. We see great potential in explicitly acknowledging the role of explanatory black boxes in the science classroom.

While this chapter focuses on recognizing the existence of black boxes and reasoning about their inclusion, fully incorporating them into teaching involves more than acknowledgment. Future research could explore additional dimensions of black box explication, such as helping students develop strategies for selecting appropriate levels of detail, teaching them to articulate their reasoning for black boxing specific details, and examining the role of black boxes in fostering scientific inquiry.

Encouraging students to critically assess and articulate the purpose of both unpacking and black boxing can lead to a deeper appreciation of the complexities inherent in scientific practices. This approach aligns with evolving pedagogical strategies aimed at nurturing a more sophisticated and reflective understanding of science.

7.5.1 Limitations

This study is based on a relatively small sample of 10th-grade biotechnology students. We acknowledge the limitations of generalizing these findings, suggesting our data as implying a potential for an explicated black box discourse in the classroom. We note the possibility that our data results from an extensive explication of the 'black-box' term to the students through the online unit and teacher's actions. Such pedagogical discourse, while within teachers' achievement, is not necessarily a natural discourse in current classrooms.

We also acknowledge the possible effect of the disciplinary context students engaged with on their ability to identify and discuss black boxes. Different disciplines might focus on different types of explanations, limiting or expanding students' engagement with mechanistic epistemology. For example, in chemistry, students might focus more on general laws and regularities (Alameh et al., 2023), while in physics they might focus on describing patterns (Tang et al., 2020). Focusing on other phenomena and disciplines in future research could strengthen the generalizability of our findings.

Our data was based on the learning products produced by students. However, we recognize that our approach did not include a thorough exploration of their thinking processes. Our presence in the classroom allowed only for limited and less systematic observations of students' discussions during their work. Implementing think-aloud protocols or interviews in the future might elucidate students' reasoning behind their decisions to either unpack or encapsulate them.

Another limitation of this study is the variability in students' prior knowledge, especially for non-cancer diseases. While about half the posters focused on such topics, students' familiarity with these subjects was not assessed. For instance, a student creating a poster on diabetes, herself diabetic, likely had extensive prior knowledge. This variability may have influenced students' ability to identify and use black boxes. Future research should systematically assess prior knowledge to clarify its role in students' use of black boxes and its broader impact on mechanistic reasoning.

Acknowledgments We thank Anat Kaplan and her students for their invaluable participation in this study. Their dedication to the learning process was essential to the success of our research. We are particularly grateful to Mrs. Kaplan for her unwavering support to foster a collaborative learning environment, which greatly enriched the study. This study was supported by the Moross Integrated Cancer Center (MICC) at the Weizmann Institute of Science.

References

Abd-El-Khalick, F. (2013). Teaching with and about nature of science, and science teacher knowledge domains. *Science & Education, 22*(9), 2087–2107. https://doi.org/10.1007/s11191-012-9520-2

Alameh, S., Abd-El-Khalick, F., & Brown, D. (2023). The nature of scientific explanation: Examining the perceptions of the nature, quality, and "goodness" of explanation among college students, science teachers, and scientists. *Journal of Research in Science Teaching, 60*(1), 100–135. https://doi.org/10.1002/tea.21792

Bachtiar, R. W., Meulenbroeks, R. F. G., & van Joolingen, W. R. (2022). Mechanistic reasoning in science education: A literature review. *Eurasia Journal of Mathematics, Science and Technology Education, 18*(11). https://doi.org/10.29333/ejmste/12512

Băetu, T. M. (2015). The completeness of mechanistic explanations. *Philosophy of Science, 82*(5), 775–786. https://doi.org/10.1086/683279

Bechtel, W. (2007). Naturalism and mechanism: Outlines of a new philosophy of science. In *Mental mechanisms* (pp. 1–48). Psychology Press. https://www.taylorfrancis.com/books/9781136679711

Berland, L. K., & Hammer, D. (2012). Framing for scientific argumentation. *Journal of Research in Science Teaching, 49*(1), 68–94. https://doi.org/10.1002/tea.20446

Berland, L. K., Schwarz, C. V., Krist, C., Kenyon, L., Lo, A. S., & Reiser, B. J. (2016). Epistemologies in practice: Making scientific practices meaningful for students. *Journal of Research in Science Teaching, 53*(7), 1082–1112. https://doi.org/10.1002/tea.21257

Cooper, M. M., & Noyes, K. (2025). Design principles for developing and analyzing formative assessments to support mechanistic reasoning. In *Understanding phenomena – Mechanistic reasoning in science education.*

Craver, C. F., & Darden, L. (2013). *In search of mechanisms: discoveries across the life sciences.* The University of Chicago Press.

de Andrade, V., Shwartz, Y., Freire, S., & Baptista, M. (2022). Students' mechanistic reasoning in practice: Enabling functions of drawing, gestures and talk. *Science Education, 106*(1), 199–225. https://doi.org/10.1002/sce.21685

Dickes, A. C., Sengupta, P., Farris, A. V., & Basu, S. (2016). Development of mechanistic reasoning and multilevel explanations of ecology in third grade using agent-based models. *Science Education, 100*(4), 734–776. https://doi.org/10.1002/sce.21217

Duncan, R. G., & Tseng, K. A. (2011). Designing project-based instruction to foster generative and mechanistic understandings in genetics. *Science Education, 95*(1), 21–56. https://doi.org/10.1002/sce.20407

Franovic, C. G.-C., Williams, N. R., Noyes, K., Klymkowsky, M. W., & Cooper, M. M. (2023). How do instructors explain the mechanism by which ATP drives unfavorable processes? *CBE—Life Sciences Education, 22*(4). https://doi.org/10.1187/cbe.23-05-0071

Fuhrmann, T., Rosenbaum, L., Wagh, A., Eloy, A., Wolf, J., Blikstein, P., & Wilkerson, M. (2024). Right but wrong: How students' mechanistic reasoning and conceptual understandings shift when designing agent-based models using data. *Science Education.* https://doi.org/10.1002/sce.21890

Gericke, N., & Mc Ewen, B. (2023). Defining epigenetic literacy: How to integrate epigenetics into the biology curriculum. *Journal of Research in Science Teaching, 60*(10), 2216–2254. https://doi.org/10.1002/tea.21856

Haskel-Ittah, M. (2023). Explanatory black boxes and mechanistic reasoning. *Journal of Research in Science Teaching, 60*(4), 915–933. https://doi.org/10.1002/tea.21817

Haskel-Ittah, M., & Yarden, A. (2018). Students' conception of genetic phenomena and its effect on their ability to understand the underlying mechanism. *CBE—Life Sciences Education, 17*(3), ar36. https://doi.org/10.1187/cbe.18-01-0014

Haskel-Ittah, M., Duncan, R. G., & Yarden, A. (2020). Students' understanding of the dynamic nature of genetics: Characterizing undergraduates' explanations for interaction between genetics and environment. *CBE Life Sciences Education, 19*(3), 1–13. https://doi.org/10.1187/cbe.19-11-0221

Illari, P. M. K., & Williamson, J. (2012). What is a mechanism? Thinking about mechanisms across the sciences. *European Journal for Philosophy of Science, 2*(1), 119–135. https://doi.org/10.1007/s13194-011-0038-2

Jahanifar, M., & Masnavi, A. (2023). Hybrid learning environment: The effectiveness of simultaneous use of real and virtual experiments on students' scientific reasoning skill. *Journal of Research in Teaching, 11*(2).

Kapon, S. (2017). Unpacking sensemaking. *Science Education, 101*(1), 165–198. https://doi.org/10.1002/sce.21248

Keil, F. C. (2019). How do partial understandings work? In S. R. Grimm (Ed.), *Varieties of understanding* (pp. 191–208). Oxford University Press. https://doi.org/10.1093/oso/9780190860974.003.0010

Krist, C. (2020). Examining how classroom communities developed practice-based epistemologies for science through analysis of longitudinal video data. *Journal of Educational Psychology, 112*(3), 420–443. https://doi.org/10.1037/edu0000417

Krist, C., Schwarz, C. V., & Reiser, B. J. (2018). Identifying essential epistemic heuristics for guiding mechanistic reasoning in science learning. *Journal of the Learning Sciences*, 1–46. https://doi.org/10.1080/10508406.2018.1510404

Livni Alcasid, G. A., & Haskel-Ittah, M. (2022). A lighter shade of black boxes: Students' interpretations of the distinction between black box explanations and unpacked mechanistic explanations. In C. A. Chinn, E. Tan, C. Chan, & Y. Kali (Eds.), *Proceedings of the 16th international conference of the learning sciences – ICLS 2022* (pp. 43–50). International Society of the Learning Sciences. https://doi.org/10.22318/icls2022.43

Livni Alcasid, G. A., & Haskel-Ittah, M. (2024). "Black boxes, full of them": Biology teachers' perception of the role of explanatory black boxes in their classroom. *Research in Science Education.* https://doi.org/10.1007/s11165-024-10191-4

Lockhart, K. L., Chuey, A., Kerr, S., & Keil, F. C. (2019). The privileged status of knowing mechanistic information: An early epistemic bias. *Child Development, 90*(5), 1772–1788. https://doi.org/10.1111/cdev.13246

Machamer, P., Darden, L., & Craver, C. F. (2000). Thinking about mechanisms. *Philosophy of Science, 67*(1), 1–25. https://doi.org/10.1086/392759

Miller, E., Manz, E., Russ, R., Stroupe, D., & Berland, L. (2018). Addressing the epistemic elephant in the room: Epistemic agency and the next generation science standards. *Journal of Research in Science Teaching, 55*(7), 1053–1075. https://doi.org/10.1002/tea.21459

Nathan, M. J. (2021). *Black boxes: How science turns ignorance into knowledge.* Oxford University Press. https://doi.org/10.1093/oso/9780190095482.001.0001

Olsher, G., & Dreyfus, A. (1999). Biotechnologies as a context for enhancing junior high-school students' ability to ask meaningful questions about abstract biological processes. *International Journal of Science Education, 21*(2), 137–153. https://doi.org/10.1080/095006999290750

Russ, R. S., Scherr, R. E., Hammer, D., & Mikeska, J. (2008). Recognizing mechanistic reasoning in student scientific inquiry: A framework for discourse analysis developed from philosophy of science. *Science Education, 92*(3), 499–525. https://doi.org/10.1002/sce.20264

Schwartz, R. S., Lederman, N. G., & Crawford, B. A. (2004). Developing views of nature of science in an authentic context: An explicit approach to bridging the gap between nature of science and scientific inquiry. *Science Education, 88*(4), 610–645. https://doi.org/10.1002/sce.10128

Talanquer, V. (2010). Exploring dominant types of explanations built by general chemistry students. *International Journal of Science Education, 32*(18), 2393–2412. https://doi.org/10.1080/09500690903369662

Talanquer, V., & Szozda, A. R. (2024). An educational framework for teaching chemistry using a systems thinking approach. *Journal of Chemical Education, 101*(5), 1785–1792). American Chemical Society. https://doi.org/10.1021/acs.jchemed.4c00216

Tang, X., & Hammer, D. (2024). "I think of it that way and it helps me understand": Anthropomorphism in elementary students' mechanistic stories. *Science Education, 108*(3), 661–679. https://doi.org/10.1002/sce.21851

Tang, X., Elby, A., & Hammer, D. (2020). The tension between pattern-seeking and mechanistic reasoning in explanation construction: A case from Chinese elementary science classroom. *Science Education, 104*(6), 1071–1099. https://doi.org/10.1002/sce.21594

Weinberg, P. J. (2018). Assessing mechanistic reasoning: Supporting systems tracing. *Journal of Pre-College Engineering Education Research, 9*(1), 3. https://doi.org/10.7771/2157-9288.1182

Chapter 8
Promoting Students' Mechanistic Reasoning in Physics Through Constructing Stop-Motion Animations: What to Consider?

Rayendra Wahyu Bachtiar, Ralph F. G. Meulenbroeks, and Wouter R. van Joolingen

8.1 Introduction

There is ample evidence (e.g., Ainsworth et al., 2011; Prain & Tytler, 2012; Tytler et al., 2020; Van Meter & Garner, 2005) that science students benefit from constructing models, which refer to a physical representation of the observed phenomenon that includes the underlying mechanism of the phenomenon (Louca & Zacharia, 2012). We address one type of dynamic models, i.e., stop-motion animation (SMA), a form of animation made up of a series of individual images (Bachtiar et al., 2021). Learning science by creating SMA has been positively evaluated in a number of studies (Farrokhnia et al., 2020). Benefits of student-generated SMAs that have been found are fostering conceptual understanding (Mills et al., 2019; Nielsen et al., 2020a), encouraging to engage with science content (Hoban, 2020; Orraryd & Tibell, 2021), promoting discussions (Mills et al., 2019), and facilitating the construction of scientific explanations (Berg et al., 2019; Nielsen et al., 2020b). Creating SMAs as a learning approach promoted students' interest in learning science and geology (Mills et al., 2020), advanced their communication skills (Deaton et al., 2013; Fridberg & Redfors, 2018) and increased intrinsic motivation for science (Kamp & Deaton, 2013; Vratulis et al., 2011).

Although SMA construction activities have advantages, its implementation in a classroom setting has proved to be challenging. Wishart (2017) pointed to the need for the appropriate preparations, e.g., making sure that the SMA software works

R. W. Bachtiar (✉)
Faculty of Teacher Training and Education, University of Jember, Jember, Indonesia
e-mail: rayendra_fkip@unej.ac.id

R. F. G. Meulenbroeks · W. R. van Joolingen
Freudenthal Institute, Utrecht University, Utrecht, The Netherlands

N. Graulich et al. (eds.), *Exploring Mechanistic Reasoning in Science Education*, Contributions from Science Education Research 17,
https://doi.org/10.1007/978-3-032-10246-1_8

properly, and students know how to use it. Furthermore, studies point out that constructing an SMA is a time-consuming process (Berg et al., 2019; Orraryd & Tibell, 2021). Berg et al. (2019) found that students spent a lot of time and effort on the SMA constructed as part of their practical work. Mills et al. (2019) pointed out that, compared to adults such as university students who are more autonomous learners, secondary school students need more teacher support during the construction of an SMA. Nielsen and Hoban (2015) suggested comparing students' SMA creation with correct expert-constructed models to scaffold students to build scientific knowledge.

As an effort to deal with these challenges, we pay particular attention to the fundamental aspects of SMA construction activities, i.e., chunking and sequencing (Bachtiar et al., 2021). In particular, we seek to understand whether and how these activities support students in developing mechanistic reasoning (MR), a type of reasoning specifying the underlying mechanisms of physical phenomena. Bachtiar et al. (2024) proposed a theoretical framework that illustrates how the cognitive process of chunking and sequencing elicits the elements of MR. As this framework centers on theoretical perspectives, practical evidence is needed to fully understand the extent to which chunking and sequencing promote the elements of MR. To this end, our current study is guided by the following research questions (RQs):

- RQ1: How do students represent an electrostatic phenomenon in an SMA?
- RQ2: How might the characteristics of students' SMA associate with their MR?

8.2 Reasoning Mechanistically Through SMA Construction

We draw on seven elements of MR proposed by Russ et al. (2008): (1) describing the target phenomenon (Element 1; E1), (2) identifying setup conditions (Element 2; E2), (3) identifying entities (Element 3; E3), (4) identifying activities of entities (Element 4; E4), (5) identifying properties of entities (Element 5; E5), (6) identifying organization of entities (Element 6; E6), and (7) chaining (Element 7; E7) (see Russ et al., 2008, p. 512–513, for the detail descriptions). Our study defines students demonstrating MR as those who provide mechanistic explanations of an observed phenomenon by describing both entities and activities (Bachtiar et al., 2022). In addition, MR includes entities at (at least) one scale level below the scale level of a target phenomenon (Bachtiar et al., 2022; Krist et al., 2019). For example, MR about the behavior of ideal gases necessarily involves descriptions of atomic particles (as entities) and the movement of the particles (activities of the entity).

Stop-motion animation (SMA) is a form of animation created by capturing a sequence of individual images, which we refer to as frames in our study. We argue that constructing sequential images implies cognitive processes, called chunking and sequencing; see the framework of cognitive processes underlying SMA construction in the study of Bachtiar et al. (2024). When the construction of an SMA was intended to model an observed phenomenon, a cognitive process refers to MR. We argue that constructing sequential frames encourages students to think in terms of two basic elements of MR, i.e., entities and activities of entities (Bachtiar et al., 2022). Additionally, other mechanistic elements may be incorporated to create a coherent sequence.

8.3 Methods

8.3.1 Research Context and Participants

Our present study aims to explore student-constructed SMAs and its association with their MR. This study was conducted in a science classroom involving tenth-grade students (aged 14–16 years) from three different classes and two different teachers at an urban international school in the Netherlands. Given the age of the students, informed consent from the parents and the students was obtained for video and audio recordings. Both teachers had more than 10 years of physics teaching experience. The first author had meetings with the teachers to introduce the study and set up the physics lesson on static electricity, in the classroom. The first author acting as a researcher and the teachers worked together during the lesson, with very different tasks. The teachers were in charge of the lesson plan, including the introduction of basic knowledge on static electricity to students. The researcher only provided guidance in the process of constructing an SMA.

8.3.2 Research Design and Data Collection

The study was conducted during three 90-minute-lessons. Table 8.1 provides an overview of the lesson plan. The followings are the descriptions of six stages of the lesson:

Stage 1: Demonstration

> In this stage, students were divided into groups of two or three. They worked together to conduct a simple electrostatic experiment, i.e., after rubbing the balloon against the hair and then bringing it near small pieces of paper, some pieces are attracted and stick to the balloon.

Stage 2: Pre-test

> In this stage, the students were given individual assignments to write answers to a question related to the experiment they had just observed. The question is "Why is the paper attracted to the balloon after the balloon has been rubbed on the hair?"

Table 8.1 The six stages of the lesson

Stages	The lesson activities	Data collected
1. Demonstration	A group of two or three students do a simple electrostatic experiment, guided by the teacher	–
2. Pre-test	Students individually answer questions about the subject	41 student pre-tests
3. Creating an SMA	Students work together (20 groups in total) to make a storyboard and create an SMA	20 artifacts containing a storyboard and an SMA for each
4. Whole-class discussion	One volunteer group presents an SMA; the other groups respond with their observations and comments	–
5. Post-test	Students individually answer questions relating to a video about electrostatics	41 student post-tests
6. Transfer test	Students individually answer a further question relating to the video about electrostatics	41 student transfer tests

Stage 3: Creating an SMA

The students were instructed to work in groups during this stage (20 groups in total). They were instructed to share and discuss their individual written answers until they reached a consensus. Afterward, they were tasked to present their collective answer in the form of an SMA.

The students were asked to start with creating a storyboard. Storyboarding is the pre-production process of an SMA, allowing students to outline a story as a visual blueprint for the entire story presented in the planned SMA. The students used a sketch in a square blank "frame", and provided the written descriptions of the drawings.

After a storyboard was complete, students used Stop Motion StudioTM to make an animation product. Using Stop Motion StudioTM, students were tasked to take digital photographs; they included annotations in these photographs, and recorded a narration to the final SMA. Note that neither the teacher nor the researcher provided any conceptual assistance at this stage, only technical support. At the end of the second stage, in total, 20 storyboards and 20 animations were collected.

Stage 4: Whole-class discussion

A whole-class discussion was held for all groups to give and receive peer feedback. This step was conducted by asking a group to present their animation and then letting the other groups respond to the animation. Note that not all SMAs were shared in this way, only one, and on a voluntary basis.

Stage 5: Post-test

In this stage, the teacher played a video and then paused it at a specific time frame to show that the balloon is being rubbed against someone's hair.

Afterward, the students worked individually to answer the question: "after rubbing the balloon against the hair, will the balloon stick to the wall? Why?" This question is considered the post-test because the questions in pre- and post- tests are essentially identical.

Stage 6: Transfer test

Once all students finished their post-test, the teacher went on playing the video (that was being played in Stage 5) to the end. This video shows that the balloon sticks to the wall after it has been rubbed. Then, the students were asked to provide the written answers to a further question, i.e., "If we do not do anything to the balloon, will the balloon fall down or keep sticking to the wall? Why?" We assigned this question to a transfer test because this test requires students to recall knowledge and skills and to apply them in new situations (Dori & Sasson, 2013).

At the end, 41 students' written answers to the post- and transfer test were collected.

8.3.3 Data Analysis

Data sources included in the analysis were 41 students' written responses to pre-, post-, and transfer tests, and 20 SMAs. Below are the details of the analysis method.

To address RQ1, the first author examined the SMAs in terms of presentation and content. For presentation, we recorded: (1) duration, (2) number of frames, and (3) presence of narration and annotations. The content analysis focused on how an electrostatic phenomenon of the balloon rubbed against someone's hair was visualized in the students' SMAs. Two criteria guided the fine-grained analysis: (1) did an SMA include depiction of microscopic particles? And (2) were frames organized in chunks representing the change of specific states at microscopic levels? Chunks were defined as adjacent frames depicting a single state of a phenomenon, with transitions labeled explicit if they clearly described changes between states (e.g., "electrons transferred into the balloon").

For RQ2, we first coded the students' written responses based on the elements of MR (as per Russ' et al., (2006) scheme). The identified codes were then discussed and refined by all the authors. After a consensus was reached, the second coder checked about 20% of the codes. Interrater reliability analysis (Cohen's kappa) resulted in 0.76, indicating near-strong agreement (McHugh, 2012). Second, the analysis was focused on categorizing MR/non-MR students. By drawing on the notion of MR in the Bachtiar et al. (2022), we labelled the students' responses as MR or non-MR based on the presence of two elements of MR, i.e., entities and activities of entities. Lastly, we cross-referenced MR and non-MR students with the content analysis of SMA to identify the association between MR and SMA. This association was further examined using the Chi-squared test.

The study sought to explore how an SMA was constructed by the students. Thus, the analysis of SMA was intended to find out how an SMA was formed, and then whether such the form might be associated with students' MR. Thus, we did not examine the SMAs for the elements of MR; rather we analyzed the presentation of the SMAs. Furthermore, to achieve this, the analysis in terms of "chunks" was guided by the framework proposed by Bachtiar et al. (2024) which addresses the construction nature of SMA.

8.4 Results

8.4.1 The Characteristics of Student-Generated SMAs About Electrostatic Phenomena

This section presents the results of analyzing 20 groups of students' SMAs regarding their presentation and content. Concerning the results from analyzing the presentation, the SMAs are different in time length, ranging from 4s (Group X5) to 28s (Group Y2); See appendices for the summary of the results. An SMA was built from a number of frames that are equal to the time length (e.g., Group X1), or more than the time length (e.g. Group X3); scc Fig. 8.1 as an example of a frame. Twelve groups (out of 20) added an annotation to the frames and 17 groups (out of 20) inserted a narration in their SMA.

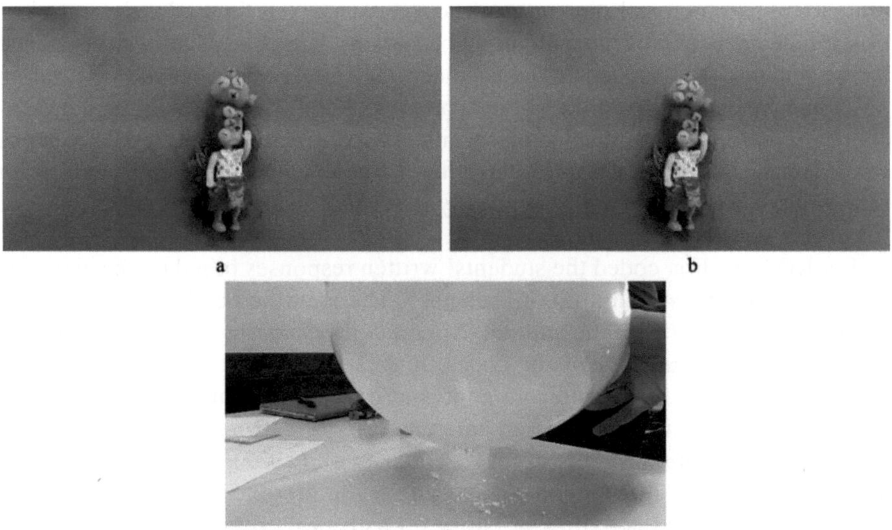

Fig. 8.1 Screenshots of the frames in the SMA made by Group Z1 (**a**, **b**) and Group Y2 (**c**); see Supplementary Animation for the animations

Table 8.2 Cross-case tabulation between SMA meeting criteria 1 and 2; Note that the cells with numbers indicate the number of student groups, whether or not their SMA design met the criteria; A total of 20 groups of students were analyzed

	Meeting criteria 2	Not meeting criteria 2
Meeting criteria 1	9	10
Not meeting criteria 1	0	1

Table 8.2 presents the cross-case tabulation resulting from analyzing the content of SMAs. The content analysis focused on two criteria. Criteria 1 concerns whether SMA presents microscopic particles, and Criteria 2 refers to whether SMA depicts the state change at the microscopic levels. For example, the SMA created by Group Z1 (Fig. 8.1a, b), includes atomic particles (Criteria 1) and illustrates the change in the states of the balloon and the hair at microscopic levels (Criteria 2). The first represents the initial state of both the balloon and the hair at the microscopic level, that is, there are an equal number of electrons and protons in either the balloon or the hair (see Fig. 8.1a). The second represents the new state, the balloon had more electrons than protons while the number of electrons in the hair decreased (see Fig. 8.1b).

The results showed that the SMA in 19 out 20 groups (e.g., Group Z1) illustrated microscopic particles, e.g., electrons; see Fig. 8.1a, b as an example. Among these 19 groups, 18 used the small pieces of clay as a model for such particles. In one of these 19 groups (i.e., Group Y1), the depiction of atomic particles was described using narration instead of visual displays. One SMA out of 20 groups, i.e., Group Y2, did not illustrate such atomic particles; see Fig. 8.1c.

Concerning the cross-case analysis, as shown in Table 8.2, an SMA in 9 groups met both Criteria 1 and 2. Ten groups created an SMA meeting only Criterion 1. In the rest of the groups, their SMA did not meet either criterion. See Appendices for the results of content analysis of SMAs by each group.

8.4.2 *MR in Pre-, Post- and Transfer Test with Respect to the Creation of SMA*

Figure 8.2 presents the total number of students involving elements of MR in the written answers of pre-, post- and transfer test. Looking at the basic elements of MR, 39, 41, and 40 students' answers of pre-, post-, and transfer test, respectively, exhibited Element 3 (entities). Also, Element 4 (activities of entities) was identified in 23, 27, and 24 students' answers of pre-, post-, and transfer test, respectively. However, no clear patterns are visible in the results of these tests.

The following Tables 8.3 and 8.4, focusing on Phase A and Phase B, respectively, present the results of the cross-case tabulation analysis intended to investigate possible associations between the quality of SMA and MR. The quality of SMA refers to an SMA meeting both Criteria 1 and 2.

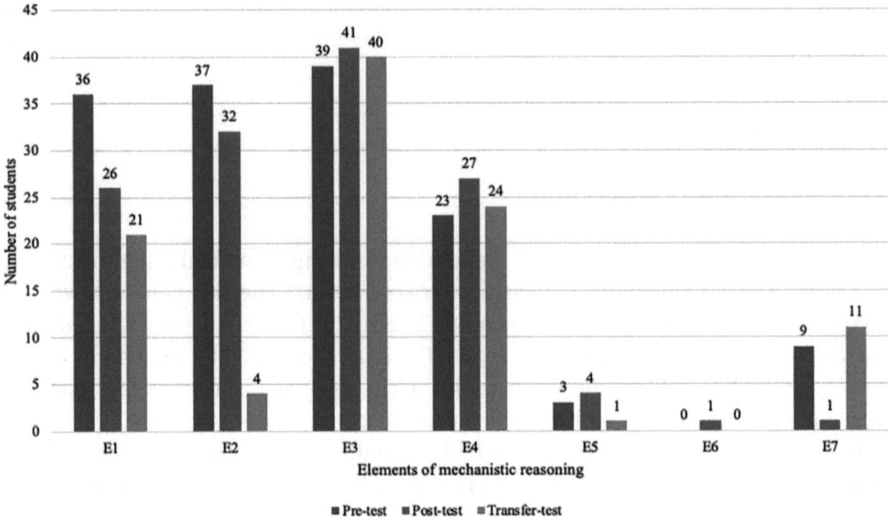

Fig. 8.2 The elements of MR identified in students' written answers; E1 to E7 refer to the terms of the elements of MR in Russ et al. (2008)

Table 8.3 Cross-case tabulation between the design of SMA and students' MR in the post-test. (Note that the cells with numbers in Tables 8.3 and 8.4 indicate the number of students)

	Students demonstrating MR	Students not demonstrating MR
SMA meeting both criteria	10	9
SMA meeting one of the criteria or none at all	2	20

Table 8.4 Cross-case tabulation between the design of SMA and students' MR in the transfer-test. (Note that the cells with numbers in Tables 8.3 and 8.4 indicate the number of students)

	Students demonstrating MR	Students not demonstrating MR
SMA meeting both criteria	12	7
SMA meeting one of the criteria or none at all	12	10

As shown in Table 8.3, among 19 students whose SMA met both criteria, there are 10 students exhibited MR in the post-test. Also, more students did not demonstrate MR (20/12) when their SMA did not meet these criteria. The Chi square test is significant ($X^2(2) = 9336$; $p = 0.0022 < 0.05$), thus implying that there is association between the quality of SMA and MR in the post-test.

Based on Table 8.4, there are ten students whose SMA in both phases met two criteria. Among these ten students, eight students' reasoning in the transfer test was identified as MR. Among 20 students whose design SMA in both phases did not meet the criteria, half of them demonstrated MR in the transfer test.

Based on Table 8.4, among 19 students whose SMA met the criteria, 12 demonstrated MR in the transfer test. The Chi square test is not significant ($X^2(1) = 0.3117$; $p = 0.56 > 0.05$), meaning that there is no association between the quality of SMA and MR in the transfer test.

8.5 Discussion and Conclusion

8.5.1 *RQ1: What Are the Characteristics in Terms of Presentation and Content of SMAs Created by Students in a Science Class?*

The first research question was addressed by analyzing SMAs in terms of presentation and content. In terms of the presentation, students' SMAs show a large diversity in terms of number of frames (4–104 for the SMA), and corresponding annotation or narration.

In terms of the content analysis, all SMAs depicted electrostatic phenomena and addressed two phases of the experiment they are supposed to model, i.e., when rubbing a balloon against someone's hair and bringing the balloon towards the small pieces of paper. The majority of SMAs (19 out of 20 groups) did include such microscopic particles. The SMAs in some of the groups depicts the change in the state of microscopic particles. The depiction is intended to specify why and how the change could happen.

As exemplified, an SMA illustrates the change of the number of electrons and protons in the balloon. The first state presents an equal number of electrons and protons in the balloon. The next state represented that the number of electrons and protons was no longer equal. A narration was also added to this SMA by stating that "the electrons from the hair transfer to the hair so that makes the balloon negatively charged."

8.5.2 *RQ2: How Does the Creation of SMA Contribute to Students' MR About Electrostatic Phenomena?*

The findings indicate that the quality of SMA is associated with students' MR. The quality of SMA was related to a specific design, i.e., (1) the inclusion of microscopic particles, and (2) an SMA depicting the change of a state at microscopic levels.

The framework we developed could demonstrate why and how the nature of SMA construction, i.e., chunking and sequencing, supports MR (Bachtiar et al., 2024). In this framework, both cognitive processes underlying chunking and sequencing (as the nature of SMA construction) promote the elements of MR, specifically entities and activities of entities.

Our current findings provide a new insight into our framework. While the framework illustrates that the nature of SMA construction elicit the elements of MR, the findings emphasize that SMA creation should depict the change of a state at microscopic levels. Our findings also add new knowledge to the literature exploring the role of SMA construction activities in supporting MR. While the studies, e.g., Wilkerson et al. (2018) and Bachtiar et al. (2021), provided empirical evidence that engaging students in constructing MR stimulated them to use MR, our current study adds that the way an SMA is constructed under a certain design also matters. Another study, by Krist et al. (2019), points out that to guide students' construction of MR, students should be directed to think at scale levels below the scale level of a target phenomenon, and to identify and unpack the relevant factors at such levels. Our current study has shown that students indeed engaged in thinking at such levels when constructing an SMA model of electrostatic phenomena.

Science education scholars (e.g., Braaten & Windschitl, 2011; Clement, 2013; Mathayas et al., 2021) regard such an SMA model as "explanatory models," a kind of scientific model visualizing the activities of unseen entities to explain why and how an observable phenomenon occurs. Explanatory models are valuable for students "provide explanations for different phenomena across many contexts" (Braaten & Windschitl, 2011; p. 649).

The findings in our current study showed that the quality of SMA exhibiting such explanatory models was associated with students' MR in the post-test, but there was no association between the quality of SMA and students' MR in the transfer test.

One possible explanation for these findings could be attributed to what Dori and Sasson (2013) call students' transfer skills, as "the ability to apply cognitive gains from one learning situation to another learning situation" (p. 369). According to Dori and Sasson (2013), the tasks in the post-test and in the transfer-test could be considered "near transfer" and "far transfer", respectively, implying that "the difficulty of a transfer task increases as we move away from" (p. 370) near transfer to far transfer. Therefore, we conjecture that for those students who did not generate MR in the transfer test, their knowledge or skills, which were acquired during the SMA construction activities, were not sufficient to construct MR in the transfer test.

8.5.3 Practical Implications

Based on the findings in this study, we make practical suggestions about what to consider when implementing SMA construction activities as a learning approach to support students' MR.

First, an SMA creation should depict the change in the state of a target phenomenon at microscopic levels. To do so, students should be encouraged to think at lower scale levels when constructing a series of frames forming an SMA. For instance, when creating the first frame considered to be the first chunk (note that one chunk can consist of one or more frames), this frame should present a state of a target phenomenon at microscopic levels. In the case of electrostatic phenomena, for instance, this first frame as Chunk 1 presents the balloon that has four electrons. When constructing the next frames as the second chunk, this second chunk represent a new state of the target phenomenon at microscopic levels, e.g., the balloon now has six electrons. Thus, the sequence from the first chunk to the second chunk depicts a change at microscopic levels in terms of an increase in the number of electrons. Therefore, this SMA not only displays a balloon that is being rubbed against someone's hair, but also represents a change in the state of the balloon at microscopic levels, i.e., additional electrons in the balloon. To support students in thinking at such lower scale levels, teachers could also provide questions, e.g., what is exactly changing from one chunk to the next chunks in terms of a state at microscopic levels?

Second, an SMA creation should provide the description of the transition from one chunk to the next chunks. Students could be tasked to provide such descriptions in the form of narration or annotation added to their SMA. As exemplified above, i.e., the sequence of two chunks represents the increase in the number of electrons in the balloon, the description of the transition between these two chunks could be: "when rubbing the balloon against the hair, electrons in the hair move to the balloon." To support students in thinking about such transition, teachers could also provide questions, e.g., why and how does the number of electrons in the balloon increase?

Finally, we suggest that when students engage in discussions, either a peer discussion during the process of constructing an SMA, or a classroom discussion intended to reflect on and discuss their SMA product, they should discuss two topics related to the particular design of an SMA i.e., whether and how (1) the sequence from one chunk to the next depicts the change in a state of a target phenomenon at microscopic levels, and (2) the transition between these chunks, which is intended to specify how and why the change happens, is described. We also suggest that teachers should facilitate such discussions, as the study by Lowell et al. (2022) showed that teacher intervention was crucial to class discussions.

8.5.4 Limitations and Future Research

We recognize limitations in our current study. The students in our study were all in one school with no students from other schools, so it might be possible that the results are biased. In our data analysis, we did not consider gender and age, thus constituting possible sources of bias.

In this study, our study involves a relatively small group of students (N = 41), which means that a larger scale study is needed for future research. We suggest that teachers should facilitate either peer or class discussions. The fact that this study was set up with minimal teacher intervention calls for further research to gain more understanding about how teachers' interventions are conducive to a productive peer or class discussion, when and in which moments teacher should intervene in discussions, thereby improving the quality of SMA.

The present study also found that students' MR was associated with a certain design of SMA. Students' MR was identified based on the presence of two elements of MR, i.e., entities and activities of entities, implying the need to explore the role of SMA in the other elements of MR. We also acknowledge that the quality of SMA could be attributed to other aspects, such as the number of frames, and frame rates, thereby the need to be explored further.

Given that the data analysis in this study did yet not consider the correctness of students' answers to the tests, we suggest that future studies explore the link between the quality of SMA and the correctness of explanations. Therefore, by gaining such an understanding, we could propose implications from such studies in terms of how the construction of SMA supports students in developing scientifically correct reasoning.

We also recommend our study should be extended to other subjects, e.g., kinematics, or electric current, so that we could gain more insights into how and under what conditions the construction of SMA contributes to developing students' MR in a broader context.

Appendices

Supplementary materials and the results from data analysis are available on OSF[1]. https://osf.io/eyrax/?view_only=78571089642a497186ba61d64581a8f0

References

Ainsworth, S., Prain, V., & Tytler, R. (2011). Drawing to learn in science. *Science, 333*(6046), 1096–1097. https://doi.org/10.1126/science.1204153

Bachtiar, R. W., Meulenbroeks, R. F., & van Joolingen, W. R. (2021). Stimulating mechanistic reasoning in physics using student-constructed stop-motion animations. *Journal of Science Education and Technology, 30*(6), 777–790. https://doi.org/10.1007/s10956-021-09918-z

Bachtiar, R. W., Meulenbroeks, R. F. G., & van Joolingen, W. R. (2022). Mechanistic reasoning in science education: A literature review. *Eurasia Journal of Mathematics, Science and Technology. Education, 18*(11), em2178. https://doi.org/10.29333/ejmste/12512

Bachtiar, R. W., Meulenbroeks, R. F. G., & van Joolingen, W. R. (2024). Understanding how student-constructed stop-motion animations promote mechanistic reasoning: A theoretical framework and empirical evidence. *Journal of Research in Science Teaching, 61*(2), 289–318. https://doi.org/10.1002/tea.21891

Berg, A., Orraryd, D., Pettersson, A. J., & Hultén, M. (2019). Representational challenges in animated chemistry: Self-generated animations as a means to encourage students' reflections on sub-micro processes in laboratory exercises. *Chemistry Education Research and Practice, 20*(4), 710–737. https://doi.org/10.1039/c8rp00288f

Braaten, M., & Windschitl, M. (2011). Working toward a stronger conceptualization of scientific explanation for science education. *Science Education, 95*(4), 639–669. https://doi.org/10.1002/sce.20449

Clement, J. J. (2013). Roles for explanatory models and analogies in conceptual change. In S. Vosniadou (Ed.), *International handbook of research on conceptual change* (2nd ed., pp. 355–371). Routledge. https://doi.org/10.4324/9780203154472-27

Deaton, C. C. M., Deaton, B. E., Ivankovic, D., & Norris, F. A. (2013). Creating stop-motion videos with iPads to support students' understanding of cell processes: "Because you have to know what you're talking about to be able to do it". *Journal of Digital Learning in Teacher Education, 30*(2), 67–73. https://doi.org/10.1080/21532974.2013.10784729

Dori, Y. J., & Sasson, I. (2013). A three-attribute transfer skills framework-part I: Establishing the model and its relation to chemical education. *Chemistry Education Research and Practice, 14*(4), 363–375. https://doi.org/10.1039/c3rp20093k

Farrokhnia, M., Meulenbroeks, R. F., & van Joolingen, W. R. (2020). Student-generated stop-motion animation in science classes: A systematic literature review. *Journal of Science Education and Technology, 29*(6), 797–812. https://doi.org/10.1007/s10956-020-09857-1

Fridberg, M., & Redfors, A. (2018). Children's collaborative learning in science Scaffolded by tablets. *International Perspectives on Early Childhood Education and Development, 22*(5), 101–115. https://doi.org/10.1007/978-981-10-6484-5_7

Hoban, G. (2020). Slowmation and blended media: Engaging students in a learning system when creating student-generated animations. In L. Unsworth (Ed.), *Learning from animations in science education* (Innovations in science education and technology) (Vol. 25, pp. 193–208). Springer. https://doi.org/10.1007/978-3-030-56047-8_8

Kamp, B. L., & Deaton, C. C. M. (2013). Move, stop, learn: Illustrating mitosis through stop-motion animation. *Science Activities: Classroom Projects and Curriculum Ideas, 50*(4), 146–153. https://doi.org/10.1080/00368121.2013.851641

Krist, C., Schwarz, C. V., & Reiser, B. J. (2019). Identifying essential epistemic heuristics for guiding mechanistic reasoning in science learning. *Journal of the Learning Sciences, 28*(2), 160–205. https://doi.org/10.1080/10508406.2018.1510404

Louca, L. T., & Zacharia, Z. C. (2012). Modeling-based learning in science education: Cognitive, metacognitive, social, material and epistemological contributions. *Educational Review, 64*(4), 471–492. https://doi.org/10.1080/00131911.2011.628748

Lowell, B. R., Cherbow, K., & McNeill, K. L. (2022). Considering discussion types to support collective sensemaking during a storyline unit. *Journal of Research in Science Teaching, 59*(2), 195–222. https://doi.org/10.1002/tea.21725

Mathayas, N., Brown, D. E., & Lindgren, R. (2021). "I got to see, and I got to be a part of it": How cued gesturing facilitates middle-school students' explanatory modeling of thermal conduction. *Journal of Research in Science Teaching, 58*(10), 1557–1589. https://doi.org/10.1002/tea.21718

McHugh, M. L. (2012). Interrater reliability: The kappa statistic. *Biochemia Medica, 22*(3), 276–282.

Mills, R., Tomas, L., & Lewthwaite, B. (2019). The impact of student-constructed animation on middle school students' learning about plate tectonics. *Journal of Science Education and Technology, 28*(2), 1–13. https://doi.org/10.1007/s10956-018-9755-z

Mills, R., Tomas, L., Whiteford, C., & Lewthwaite, B. (2020). Developing middle school students' interest in learning science and geology through slowmation. *Research in Science Education, 50*(4), 1501–1520. https://doi.org/10.1007/s11165-018-9741-8

Nielsen, W., & Hoban, G. (2015). Designing a digital teaching resource to explain phases of the moon: A case study of preservice elementary teachers making a slowmation. *Journal of Research in Science Teaching, 52*(9), 1207–1233. https://doi.org/10.1002/tea.21242

Nielsen, W., Turney, A., Georgiou, H., & Jones, P. (2020a). Creating a digital explanation in preservice teacher education: Scientific knowledge represented in a digital artefact. In L. Unsworth (Ed.), *Learning from animations in science education* (Innovations in science education and technology) (Vol. 25, pp. 229–248). Springer. https://doi.org/10.1007/978-3-030-56047-8_10

Nielsen, W., Turney, A., Georgiou, H., & Jones, P. (2020b). Working with multiple representations: Preservice teachers' decision-making to produce a digital explanation. *Learning: Research and Practice, 6*(1), 51–69. https://doi.org/10.1080/23735082.2020.1750673

Orraryd, D., & Tibell, L. A. E. (2021). What can student-generated animations tell us about students' conceptions of evolution? *Evolution: Education and Outreach, 14*(1), 1–18. https://doi.org/10.1186/s12052-021-00153-z

Prain, V., & Tytler, R. (2012). Learning through constructing representations in science: A framework of representational construction affordances. *International Journal of Science Education, 34*(17), 2751–2773. https://doi.org/10.1080/09500693.2011.626462

Russ, R. S., Scherr, R. E., Hammer, D., & Mikeska, J. (2008). Recognizing mechanistic reasoning in student scientific inquiry: A framework for discourse analysis developed from philosophy of science. *Science Education, 92*(3), 499–525. https://doi.org/10.1002/sce.20264

Tytler, R., Prain, V., Aranda, G., Ferguson, J., & Gorur, R. (2020). Drawing to reason and learn in science. *Journal of Research in Science Teaching, 57*(2), 209–231. https://doi.org/10.1002/tea.21590

Van Meter, P., & Garner, J. (2005). The promise and practice of learner-generated drawing: Literature review and synthesis. *Educational Psychology Review, 17*(4), 285–325. https://doi.org/10.1007/s10648-005-8136-3

Vratulis, V., Clarke, T., Hoban, G., & Erickson, G. (2011). Additive and disruptive pedagogies: The use of slowmation as an example of digital technology implementation. *Teaching and Teacher Education, 27*(8), 1179–1188. https://doi.org/10.1016/j.tate.2011.06.004

Wilkerson, M. H., Shareff, R., Laina, V., & Gravel, B. E. (2018). Epistemic gameplay and discovery in computational model-based inquiry activities. *Instructional Science, 46*(1), 35–60. https://doi.org/10.1007/s11251-017-9430-4

Wishart, J. (2017). Exploring how creating stop-motion animations supports student teachers in learning to teach science. *Journal of Research on Technology in Education, 49*(1–2), 88–101. https://doi.org/10.1080/15391523.2017.1291316

Part III
Processes of Constructing Mechanistic Explanations

Chapter 9
On the Productivity of Mechanistic Reasoning in Elementary Science Classroom

Xiaowei Tang, Deying Wen, and Ning Pan

9.1 Introduction

From the new mechanist perspective, the work of science is organized around the search for mechanisms (Craver & Tabery, 2015; Glennan et al., 2022; Machamer et al., 2000). Mechanistic explanations reveal the processes producing, underlying, and sustaining natural phenomena, providing us with insights about how the world works (Craver & Tabery, 2015). As science educators, we hope students can engage in mechanistic reasoning (MR hereafter), and experience how it interacts and intertwines with other scientific practices during the process of constructing mechanistic explanations (De Andrade et al., 2022; Haskel-Ittah, 2023; Krist et al., 2019).

When advocating for the importance of engaging students in MR, a question the first author frequently asked is: how do I tell *when* to open space for mechanistic reasoning? Intuitively, it seems fair to answer "When you sense your students start to do it." Yet further thought points to the need to take the question more seriously.

Constructing mechanistic explanations is a challenging and messy task (Abrams & Southerland, 2001; Haskel-Ittah & Yarden, 2017; Talanquer, 2024). It requires tentatively decomposing the target phenomena into connected activities and entities (Bechtel & Richardson, 1993) and recomposing the elements to form an explanatory account (Bechtel & Abrahamsen, 2013). The process often requires borrowing

X. Tang (✉)
Faculty of Education, University of Macau, Taipa, Macao, China

D. Wen
Jiangbei District Teachers' College of Chongqing, Jiangbei, Chongqing, China

N. Pan
Huaxin Experimental Primary School, Jiangbei, Chongqing, China

N. Graulich et al. (eds.), *Exploring Mechanistic Reasoning in Science Education*, Contributions from Science Education Research 17,
https://doi.org/10.1007/978-3-032-10246-1_9

schema from known mechanisms (Darden, 2006), manipulating the parts or the whole to test tentative conjectures (Craver, 2007), and working with accounts containing unknown gaps, or black boxes (Craver & Tabery, 2015; Haskel-Ittah, 2023). With the limited classroom time at hand, the teachers have every right to wonder whether opening space for MR would be a productive choice.

To answer the question, the present study examined a case where a group of Chinese elementary teachers made intensive efforts to initiate and facilitate MR in a fourth-grade inquiry unit. They experienced, and the first author observed, both productive and not-so-impressive moments of MR. Through analysis of these moments and theoretical reflections, we explore the following research questions:

- When is the elementary students' MR productive and how?
- What condition the productivity of the students' MR?

We hope the exploration can provide some initial ideas on how practitioners can identify (or construct) appropriate opportunities for young students to engage in MR. In the pages that follow, we consider common frameworks and criteria for understanding MR in science education, the various ways MR can contribute to explanation construction, as well as the resources needed for productive engagement in MR. The understandings gained serve as a framework that guides our case analysis.

9.2 Frameworks and Criteria for Understanding MR

In science education, the most popularly adopted concept of mechanism came from Machamer and colleagues (MDC, 2000), which defines mechanism as a system of organized entities (parts) and activities responsible for the regular change from the start to the finish conditions. Russ et al. (2008) introduced this framework and translated it into a framework for recognizing MR from classroom discourse, including a set of structural components: describing the phenomenon, identifying set-up conditions, identifying entities, identifying properties of entities, identifying activities, identifying the organization of entities, and chaining backward and forward. Building on that, Krist et al. (2019) suggested the need to think across scalar levels, identify and unpack entities, properties, and rules (usually including patterned activities) at least one level below the target phenomenon, and link them back, suggesting how they give rise to the phenomenon. This framework also identifies heuristics that guide mechanistic reasoning. Haskel-Ittah (2023) added to that the need to identify black boxes, that is, where we know a causal connection is needed, but do not have the mechanistic specifics yet (See Chap. 7 for details).

Ideas derived from these frameworks inform how we understand and evaluate students' MR. The criteria highlight the degree of difficulty and sophistication of reasoning. For instance, conjecturing what happened at a lower scalar level is considered more difficult than describing the phenomena (Russ et al., 2008; De Andrade et al., 2022), and reasoning across levels is considered a key feature of quality MR (Berg & Hulten, 2024; Keiner & Graulich, 2020). In terms of structural components, identifying activities and organizational features is considered more abstract and therefore more advanced than identifying entities (Moreira et al., 2019). Chaining and forming mechanistic accounts with fewer black boxes marked the development of explanations, and are often taken as signs of reasoning sophistication (Haskel-Ittah, 2023; Keiner & Graulich, 2020; Russ et al., 2009).

The above frameworks and criteria illustrated important features of quality MR. Yet when it comes to classroom interactions, it could be hard to consider the quality of MR simply based on what components are involved or what scalar level the reasoning gets at. Is phenomenal description always the most basic step, and therefore, contributes less than the more abstract or difficult MR elements? What if the highlighted phenomenal feature plays a critical role in selecting reasonable mechanistic accounts? And is it always more advanced to identify activities or organizational features? What if the identified details are not anchored on theoretical knowledge or evidence and can hardly be connected back to the phenomena? Within an ongoing inquiry discourse, MR is unavoidably integrated with other scientific practices (Ford, 2008) and develops in directions not fully predictable. It requires a more dynamic and contextualized perspective and a focus on understanding the roles specific MR plays in pushing forward the ongoing explanation construction process.

9.3 Productivity of MR in Explanation Construction

Our use of the term productivity aligns with the way we heard teachers use it. They hope to know when they can expect "productive" MR from the students, and by productive they usually mean that it can get the inquiry going, that the explanatory effort can move forward rather than stuck somewhere. It does not refer to the development of a canonical mechanistic account or an account involving all the structural elements, but that MR should somehow facilitate the effort of sensemaking and contribute to the ongoing process of explanation construction. Craver and Tabery (2015) characterize mechanisms as consisting of phenomenon, parts (entities), causation, and organization. Here we consider how MR in each aspect can contribute to explanation construction, and the conditions needed for MR to make such contributions.

First, a mechanism is always a mechanism of *some* phenomenon. The boundary of the phenomenon defines what needs to be explained (Craver & Tabery, 2015). The more specific we can characterize the target phenomenon, the better we can confine the level of MR and the range of mechanistic possibilities (Craver, 2007; Darden, 2006). Providing phenomenal information that leads to further confinement of mechanistic accounts is one way to contribute to the explanation construction process. To obtain such information, one would need to know what to observe and how to observe. What to observe is often driven by existing mechanistic ideas. How to observe would require specific procedural knowledge.

Second, explanation construction requires identifying entities, often from a lower scalar level (Bechtel & Richardson, 1993; Craver, 2007; Krist et al., 2019). The ideas of what may count toward entities have to come from somewhere. According to Bechtel and Richardson (1993), entities can be identified by breaking down the mechanism into potential operations and assigning the operations to components that may carry out them. These components can be material or purely theoretical (Krist et al., 2019). In some cases, with certain prior knowledge, entities can be identified from observation on a low-inference basis (e.g., knowing that neural activities consume energy and require oxygen-rich blood, one can identify from fMRI images the brain regions accounting for certain brain functions). In many other cases, entity identification starts with much less information. One may need to borrow a known schema through analogical reasoning, or reason forward or backward from some known entities (Darden, 2006). The analogy-informed identification process often involves rounds of hypothesis formation, coherence checking, and adjustments (Darden, 2006). Breaking down a mechanism and conjecturing potential entities is a major contribution to explanation construction. To do this, one needs to draw on either direct knowledge on the target phenomenon or some cognitive resources that can anchor analogical reasoning.

Third, a mechanistic explanation should illustrate causal relationships (Craver & Tabery, 2015). The involvement of causation implies activities and/or interactions, as entities cannot cause anything unless doing something. At the operational level, we believe causality should afford manipulation of the phenomenon through interventions on the entities (Woodward, 2013). We can inhibit, excite, or modify an entity, and learn about its causal role by attending to the resulting phenomenal differences (Bechtel & Richardson, 2010). One can contribute to explanation construction by conjecturing causal relationships and testing it through manipulations. To form conjectures on causality, one needs to render an unfamiliar pattern into a set of familiar patterns (Glennan et al., 2022), which requires drawing on familiar patterns informed by prior knowledge, analogies, p-prims, etc. To test it, one needs to be able to design and conduct relevant manipulations.

Finally, organizational features also play significant roles in the construction of mechanistic explanation. For one thing, phenomenal level information of spatial and temporal organization often serves as the first clues for mechanistic conjecturing (Darden, 2006). For instance, the temporal cycle of lunar phases and its shift in sky locations imply a mechanism involving circulation, relative motion, and relative positions. A fit mechanism should unify all these organizational features. For another, a mechanism often needs to illustrate organizational features that afford entities to function or interact in certain ways. Such features also need to be reasoned out through manipulations. Drawing mechanistic clues from phenomenal-level organizational features and identifying activity-affording features can both contribute to the construction of mechanistic explanations. The former often requires connecting observable features with some existing reasoning resources. The latter requires the capability to manipulate relevant features.

As a type of model, mechanisms are "inevitably partial, abstract, idealized, and plural" (p. 147, Glennan et al., 2022). The black boxes in a mechanistic account can mark where information fell short, or where more details do not matter for the current explanatory purpose (Haskel-Ittah, 2023). Selecting appropriate black boxes to unpack is key to MR productivity. Theoretically, the selection should depend on what mechanistic details are relevant to the explanatory goals (Craver, 2007). Practically, whether the participants have access to the necessary cognitive resources and manipulations constitutes a critical factor.

The understandings on how MR can contribute illustrated above guide our attention in distinguishing the productive and not-so-productive instances of student MR. When diving into the discourse for conditions affording or constraining reasoning productivity, we pay particular attention to the availability of resources and manipulations as well as the explanatory goals.

9.4 Methodology

9.4.1 The Case

In 2022, the first author offered a workshop on MR to elementary science teachers practicing within a district of a Southwestern China metropolis. She explained why MR is considered a key scientific practice, and introduced through case analysis activities Russ' (2009) framework for identifying MR from classroom discourse. A group of seven teachers became interested in opening classroom space for MR.

At an earlier time, the group leader, Mr. Shu, noticed that many fourth graders held a matter-based conception of sound (Shu et al., 2023). To challenge the conception, he designed and implemented an inquiry activity on why a string phone can only transfer sound when the string is stretched. The group then co-taught a unit by expanding on that inquiry activity. Instead of starting with a fixed design, the group allowed their teaching ideas to grow naturally. When each group member taught, all the others observed, either on spot or online. After each lesson, they held a group discussion to further understand the emergent student ideas and plan the next lesson accordingly. Together the group taught nine lessons on sound generation, sound transmission, and pitches. Our study focuses on the first six lessons that covered sound generation and transmission. Table 9.1 provides details about the instructors, the inquiry topics, and the main activities of each lesson.

All the lessons were conducted in a fourth-grade class of a local school over 4 weeks, during their regular science class time. The school got a slightly above-average ranking in a recent district-level science assessment. The class has a roughly 1:1 gender ratio (24 males and 22 females).

Table 9.1 Lessons on sound generation and transmission

Lesson	Instructor	Background	Topic	Main activities (*: involving MR
1&2	Mr. Shu	Bachelor's and master's degrees in science education; 15 years of elementary teaching experience	Why string phone can only transfer sound when the string is stretched?	*Generating initial explanations; Further explorations of the phenomenon; *Explanation evaluation and revision based on further information collected; *Predicting and testing how a "V" shape string phone works; *Predicting and testing how a wire-based string phone works
3	Ms. Huang	Bachelor's degree in science education; 16 years of elementary teaching experience	How is sound generated?	Reviewed ideas; *Discussion on what vibration is; Producing sounds with rubber bands and steel rulers; *Discussion on what produces sound, how to define vibration, and their possible relationships; Exploring vibration observation *Discussion on whether sound can occur without vibration
4 & 5	Mr. Pan	Bachelor's degree in biology; 22 years of elementary teaching experience	Is it possible to have vibration without sound?	*Reviewed understandings; *Exploring situations where vibration leads to no sound; *Students hypothesized about the string's vibration direction; *Observing vibrations transmission on a jumping rope and discussing why sound only travels on stretched strings;
			In what direction does the string vibrate?	*Discussion on whether sound and vibration transmit together; Exploring vibration direction in the string phone; *Exploring the whole process of sound generation and transmission in the string phone; Demonstrative experiment: sound vibrates a membrane
6	Ms. Yang	Bachelor's degree in textile engineering and master's degree in science education; 3 years of elementary science teaching experience	How does sound transmit into our ears?	*Discussion on how sound transmits in ears; *Exploring sound transmission in ear structure; Exploring how ear drums work; *Further discussion on how ear structure transmits sound; *Discussion on what sound is and whether the air vibrates; Demonstrative experiment: Sound in a vacuum chamber

9.4.2 Data Collection

The whole unit has been videotaped with informed consents. The first author identified episodes with students engaging in MR. She then conducted one-on-one interviews with the instructors regarding their lesson plans, their interpretations of the students' mechanistic ideas, and their rationales for how they handled the ideas. The interviews were also videotaped. Both the classroom data and the interview data were fully transcribed. Since the data was shared 1 week afterwards, we did not have the opportunity to interview the students. All the idea interpretations were therefore based on the classroom data and the teacher's personal understandings.

9.4.3 Data Analysis

The episodes were divided by natural shifts in topics. A discussion focusing on a sole topic (e.g., generating initial explanations) counts as one episode. For a discussion with topic shifts in the middle (e.g., from what produces sound to how to define vibration), the interactions around each topic constitute an episode.

We looked at MR productivity by focusing on how the reasoning work contributes to the ongoing explanation construction process. For types of contributions, we refer to the ideas identified from the literature, including informing and confining the explanatory task with phenomenal description, conjecturing and checking the involvement of potential agents, activities, possible causal relationships, and lower-level organizational features. We also attended to reappearing ideas and discrepancies, as they indicated stagnation in development and lower productivity.

We then further analyze the representative instances to illustrate how students drew on various resources and manipulations in productive MR, as well as what conditioned the not-so-productive episodes.

9.5 Results

Below we reported in themes the productive features of student MR and the constraining issues. Each theme is illustrated with typical instances.

9.5.1 Conjecturing Mechanistic Elements by Drawing on Various Resources

When constructing tentative explanations with less information than needed, students first engaged in conjecturing what entities are at work and how they possibly lead to the target phenomenon. Our case involved many such instances.

The unit started with students experiencing string phones (premade with cotton string and two paper cups) in pairs. In the sharing afterward, Mr. Shu foregrounded a phenomenon some students noticed: “We can only hear when the string is stretched; when the string is relaxed, we hear nothing.” He then drew a straight line representing the stretched-string situation and a curved line representing the relaxed-string situation. His words and drawing highlighted a piece of spatial organizational feature (straight vs. curved). Students then generated, and Mr. Shu recorded, many mechanistic conjectures for “what could have happened:”

9.5.1.1 Differential Vibrations

Shan thought the sound “has its own vibration,” which is “straighter and greater when it is a straight line”, and “curved and small” when traveling along the curved string. That leads to whether we can hear the sound. Here he identified “vibration” as the key entity that travels along the string, showing a partial understanding in his description: “greater” or “smaller” vibration makes sense but “straighter” or “curved” vibration does not. The idea likely came from some shallow popular science source that introduced vibration without fully explaining the concept. By treating observable organizational features (i.e., “straight” and “curved”) as entity features, he avoided further elaborating on the activities or causal relationships involved.

9.5.1.2 Traveling in a Straight Line and Getting Stuck

Fu conjectured that the sound “has another name called supersonic wave,” which has the feature of “only traveling in a straight line.” In response to a later challenge, she clarified that “supersonic wave” was an analogy to sound since they “behave in the same way.” Like Shan’s “vibration,” “supersonic wave” likely came from some fact-based knowledge source.

Qi followed by conjecturing the activity of sound “gets stuck in the curve and cannot go up,” causally accounting for its incapability to travel through a curved string. This idea treated sound as a matter-based entity and applied a pattern familiar to everyday experience: an inanimate object in motion (e.g. a rolling marble) would come to a stop when hitting the bottom of a curve (e.g., a pit).

Building on that, Peng conjectured “sound speed” as a key temporal organizational feature. He took sound speed as caused by speaking, that “you have to speak faster, and louder” to make sound that “travels with a very fast speed,” and assigned it a causal role in the target mechanism, that sound with “fast speed” can “go through (the curve) immediately.” The conjecture could have drawn on the common phenomenological primitive (p-prim) of *overcoming* (diSessa, 1993), or everyday experiences with speed-related motion pattern of mattered-based objects (e.g., a fast-rolling marble can hit the bottom of a pit and go out).

9.5.1.3 Traveling Time Makes a Difference

Liu conjectured "limited traveling time" as a key feature of sound. He explicitly drew on a piece of math knowledge that "a straight line is the shortest between two points," reasoning that the sound "would have to travel for a longer time" in a curved string. He further suggested that "we can't guarantee that the sound has the function to detour on its own," indicating potential alignment with "gets stuck in the curve." The "longer time" needed, when held against the feature of "limited traveling time," plays a causal role for the failure of sound transmission on the curved string.

9.5.1.4 Traveling Like a Car

Ju drew an analogy with the everyday experience that "a car travels very fast in a straight line" but "has to reduce its speed when going through a curve," conjecturing that sound "would slow down in the curve." Aligning with Liu's conjecture of "limited traveling time," Ju suggested that "sound is fading every second," as evidenced through everyday experience that sound from far away "would reduce a lot," He then proposed the slowing down in sound transmission as what causes sound to "get so weak that we feel we can't hear it."

These initial MRs are far from being canonically correct, yet epistemically productive in that they form a pool of mechanistic accounts that can be evaluated, tested, or developed. The phenomenal description activated in the students all kinds of preexisting cognitive resources (e.g., prior knowledge from formal or informal learning, everyday experience, p-prims, etc.) related to sound and the particular "stretched vs. curved" spatial organization. They served as grounds for these initial conjectures. This is the sort of practice scientists would engage in when exploring an unknown phenomenon, only that they have access to a much greater storage of cognitive resources. The conjectures demonstrated that many students applied to sound transmission the familiar motion patterns observed in regular objects. The varied accounts and the differential conceptualizations exposed effectively contextualized and motivated further inquiry.

Similar moments of free conjecture sharing occurred throughout the unit. For instance, when exploring how sound is generated, the students conjectured multiple accounts on whether vibration is necessary for sound generation. These ideas, again, prepared the ground for further explorations into what counts as vibration and how vibration relates to sound. It seems that as long as students can anchor mechanistic conjectures on *some* preexisting resources, their reasoning effort can be productive.

9.5.2 MR Through Extended Coherence Checking

MR can also occur through coherence checking, which typically integrates argumentation, focusing on demonstrating how proposed mechanistic details conflict or cohere with what one holds for sure. Complex mechanistic accounts usually provide more possibilities for coherence checking than common hypotheses.

In our case, coherence checking emerged as early as the class started to share their initial mechanistic conjectures. For instance, when Fu identified sound with "the supersonic wave" that only "travels in a straight line," Wu immediately pointed out that "supersonic wave" would be "beyond human beings' audible range." In response, Fu clarified her analogical purpose. Wu followed up right away: "Then why, when a person shouts at you around the corner, you can still hear it?" highlighting a discrepancy between the proposed feature of "traveling in a straight line" and a phenomenon from everyday experience.

More coherence checking emerged after the class further examined the phenomena and identified new phenomenal features through careful observation or manipulations. For example, many supported treating vibrations as a key entity, since it coheres with "the slight tingle on fingertip" they felt when touching a stretched string during sound transmission. Some found support for the conjecture that sound "gets stuck in the curve," as in the curved-string situation, vibration can be well felt on the speaker side of the string but greatly reduced or disappeared on the receiver side. Some students based their coherence checking on evidence from self-designed manipulation. Liu's group manipulated the loudness of the sound and the curvature of the string, attending to whether and how much sound got transmitted. If the sound "gets stuck in the curve", they reasoned that it should be able to get out more easily when it goes faster (which they associated with "louder") and when the curve is shallower. The finding that they "could hear a bit, though very weak" cohered with this mechanistic conjecture. They also detected no comparable change in sound transmission time as the loudness of the transmitted sound varied significantly, suggesting an incoherence with the third and the fourth accounts, which foreground this temporal organizational feature. Through coherence checking, the students screened the mechanistic accounts they proposed in a rational way.

Coherence checking can also be facilitated or even guided by the teachers. During the inquiry into sound generation, for instance, the class disagreed on whether all the sound-producing processes involve vibration (defined as back-and-forth motion of some object). Those who thought vibration is unnecessary suggested exemplary situations such as scratching the table with a steel ruler or clapping, where they heard sound but observed no vibration. To facilitate further reasoning, Ms. Huang provided tuning forks and water basins, asking the students to observe tuning forks' sound-producing phenomenon in "any way you can think of." The students reported "feeling a slight tingle" when touching a sound-producing tuning fork and seeing "the wrinkles on the water surface" when using it to touch the water. The resemblance between the tingle from the tuning fork and that from the sound-transmitting string allowed the students to identify water wrinkles with vibration. From here Ms. Huang extracted a general "how to observe" strategy: "When you cannot directly see the vibration, you can observe it indirectly with the assistance of other things, such as water." It opened a door for observing the less obvious vibrations:

1. Shi: Um, if we put the ruler on the table, and tap the table with hand, you can see the ruler jumping a bit.
2. Ms. Huang: Ok, great. Any other approach? Any other way to observe the vibration?

3. Liu: we found that if we scratch the table, the water surface would also show some wrinkles.
4. Ms. Huang: That works as well, right? Very smart. Oh, you have more? Go ahead.
5. Wen: (pointing at a pen on the table) We saw this rolling. Through rolling you can observe some vibrations.

The new observations and inference-making approaches further confined the phenomenon and challenged what the students took as the evidence of incoherence. As a result, the class largely accepted that the activity of vibration played an important role in all the sound-producing phenomena.

There were also situations where teachers came up with manipulations according to the students' mechanistic understanding. For example, when the class explored situations where vibrations do not produce sound, some students suggested that the difference is in whether the vibration was "a motion controlled by hand" or "a motion carried out by the object." In response, Mr. Pan placed a steel ruler between the pages of a science textbook, showing that a pluck on a long projecting part gave no sound, while a similar pluck on a short projecting part produced sound. The new phenomenon challenged the students' idea by showing how vibration not controlled by hand may or may not produce sound. One student, Fei, suggested that "the (ruler's) back part hits the surface it rests on" serves as "the source of the sound." Based on this causal relationship, he conjectured "the area (of the ruler) inside the book" as a key spatial organization feature that determines whether audible sound can be produced.

Following Fei's logic, Mr. Pan introduced another manipulation. He set the length of the ruler inside the book as the length of the book and varied the length of the projecting part (for a shorter projecting part, part of the ruler stuck out from the other end of the book). Pressing the book tightly with one hand, he then repeated the previous experiments. The unchanged outcomes suggest incoherence with Fei's conjecture.

In the above instances, besides conjecturing mechanistic accounts, student MR also contributed to explanation construction through extended coherence checking. While some coherence checking simply drew on prior knowledge, most involved further observations or manipulations, which provided data that can be checked against predictions derived from the proposed accounts. The observational and manipulative strategies provided by the instructors played significant roles in sustaining such MR effort. While it could be hard for the students to generate these strategies on their own, the way they reasoned with the data showed that they could well follow the rationales.

9.5.3 Piecing Together a Fuller Account

In the focal case, we also identified student MR effort of piecing together their mechanistic understanding and form fuller explanatory accounts.

For instance, by the end of the second lesson, most students recognized the role of vibration, or the "slight tingle." After the class unpacked vibration into "the back-and-forth motion of an object" and figured out that sound is produced when something is in vibration, Ms. Huang posed a question: "Where does the vibration on the string come from?" To that Liu offered an answer that traced back to "our vocal cords," where the vibration was first generated. Following the spatial organization of the string phone, he then identified "the first cup," "a knot in the cup," and "the string in the middle" as the entities that take on vibration successively. He used vibration and "back-and-forth motion" interchangeably, showing an understanding that it is an *activity* passing on between these entities.

Such forward chaining continued as more parallel and vertical black boxes were unpacked. In the sixth lesson, after learning about human ear structure and exploring how eardrum works, Ms. Yang asked for "a complete explanation on sound transmission using the string phone." Wu shared the following account:

> I think our ears hear sound in a similar way the string phone transmits sound. When our vocal cords vibrate and produce sound, the speaker cup of the string phone vibrates with it. Via the string the vibration comes to the receiver cup and then our ear. Then our ears hear sound in the same way. Like the auricle, it looks like the phone-, the speaker cup on the string phone. It's for collecting the sound. The external auditory canal, like the string, is responsible for transmitting the vibration. The eardrum and the ear bone function like the receiver cup. They magnify the sound vibration, and the lymph within the cochlea vibrates with it. Thc vibration-, the magnified vibration is then received by the auditory nerve and passed to the brain. The brain then interpreted it.

Here Wu drew on the string phone mechanism developed earlier as an analogical resource for modeling how ears work. The analogy started with phenomenal-level similarity in spatial organization: the auricle looks like the cup, and the external auditory canal looks like the string. Functional analogy and chaining then followed. This "complete account" covered a relatively complete sequence of vibration transmission, but did not unpack vibration into back-and-forth motion. Other students who explained the process in this discussion or later showed similar focus.

In both instances, we see students constructing complete mechanistic accounts following clues from the spatial organization, practically or analogically. The previously constructed mechanistic idea can also serve as analogical resources for later MR. Students also held a question-based measure of mechanistic completeness (Craver & Kaplan, 2020). When the focus is on "sound transmission through the string phone," the specifics of the vibration motion were considered less relevant and therefore left unpacked.

9.5.4 Prematurely Established Canonical Explanation

By the end of the sixth lesson, the students all identified vibration as "back and forth motion" and agreed on the need of high-frequency vibration in sound production, but there were still differences in their understanding. To Wu and Tai, sound *is* the back-and-forth motion vibration stands for. To Tao, sound transmits "with the assistance of vibration." To Liu, sound is a type of "microwave" that can vibrate on different objects. Whether sound and vibration are one or accompanying but separate activities (or entities) is the core discrepancy.

It was not the first time such a discrepancy emerged. Back at the start of the fourth lesson, when Mr. Pan asked "What is your understanding of sound now?" Tai and Liu provided different conjectures:

1. Tai: I understand that sound transmits in vibration.
2. Mr. Pan: Ok, that is it? Anything else? Your turn.
3. Liu: I think sound would follow a certain object-, like a string, like it would vibrate on such an object and get transmitted. We know from last lesson the mechanism of sound transmission in the string phone. There is a string, and sound would transmit to the string and vibrate. After the vibration it would go through the other cup. So I think sound is transmitted through attaching to some object and vibrating.
4. Mr. Pan: Ok, so we all think sound has something to do with vibration. In the last lesson there is another conclusion. Anyone remember it?

In contrast to Tai's canonical response, Liu conjectured sound as an entity that would "attach to some object" (activity/spatial organization) and "vibrate" (activity) (line 3). Mr. Pan dismissed the difference by focusing on their common ground ("we all think sound has something to do with vibration," line 4). The later conversation refocused on an issue Mr. Pan tried to solve in this lesson, that "there are situations where we can observe the vibration but hear no sound." This predetermined goal kept him from attending to differences between these ideas.

The conflict came back in the fifth lesson when a student, Ju, considered the possibility that vibration may continue transmitting even after the sound fades. Mr. Pan tossed the idea back to the class, and the responses revealed various conjectures in this regard. Tai thought sound gets into the ear before vibration, whereas Wen took the "vibration without sound" situation as evidence that "vibration goes first." Mr. Pan did not probe into the reasoning behind the conjectures. Instead, he hit the tuning fork on the table and asked a student to pinch it right away, highlighting that the sound disappeared immediately when vibration stopped. He then established a conclusion ("if the vibration is transmitted to the other side of the string phone, the sound should also be transmitted.") and dropped the topic.

Mr. Pan's manipulation showed how sound and vibration occur and disappear together, yet never touched the core mechanistic discrepancy of whether the two are one or separate, entities or activities. Elementary-level canonical expression that sound is produced when an object vibrates and transmits through vibration can be easily interpreted as describing sound as an *entity* produced and transmitted through the *activity* of vibration. When canonical knowledge was established without fully addressing the discrepancies in students' mechanistic conjectures, it can block students' MR from contributing to the ongoing explanation construction process. The unsolved discrepancy could also come back again and again and baffle the related reasoning effort.

9.5.5 When Reasoning Resources Are Scarce

The class spent about fifty minutes exploring the direction of sound vibration on the string phone. The teacher eventually guided the class to the conclusion, as the students showed great difficulty reasoning on this specific topic.

The exploration started after the class agreed that vibration frequency matters for sound generation. A student, Liu, updated the explanatory account on why curved string cannot transmit sound, replacing "get stuck in the curve" with a new conjecture: vibration would "slow down" and becoming "lower in frequency" in the curve. The MR effort reasonably integrated the empirical finding that links low frequency with no sound. When Mr. Pan asked "How do you think the (stretched) string vibrates?" The class responded with "in high frequency," indicating an alignment with Liu's idea. Mr. Pan rephrased the question into "How about the direction? The vibration direction?" This sudden shift toward a spatial organizational feature never mentioned before caused a 10-second pause.

When students finally started to share, they sounded unconfident. Only a few students gestured an "up-and-down" guess. Lei then shared with multiple expressions of uncertainty: "I'm just guessing, maybe, probably it's vibrating in multiple directions [gesturing various directions vertical to the string]." Shi added that the vibration "may not distribute evenly" and linked it to sound variation, offering no further rationale. None of the conjectures came with any explicit evidence or reasoning.

Mr. Pan introduced two activities exploring the vibration direction. First, he demonstrated with a jumping rope that hand-controlled vibrations in multiple directions (vertical to the string) can pass through either stretched or curved rope, hoping that the incoherence in transmission patterns can challenge the students' guesses. The class, however, had difficulties reasoning about this simulation. Most students denied the analogy, either suggesting that the hand-controlled vibration on the rope was "much greater than the vibration on the string," or separating sound and vibration: "vibration can go up and down, but when sound is climbing up it will gradually disappear." This latter reasoning coheres with the simulated phenomenon, but contradicts the phenomenal features identified earlier—that the vibration cannot pass to the other side of the curve.

Second, Mr. Pan gave each student a paper cup with a small Styrofoam ball hanging from the upper edge of its bottom. Students were instructed to speak into the cup and observe the vibration of the ball. Among the six students who shared their observational results, only two described the ball's motion as "to-and-fro." Two stated that the ball "vibrated in multiple angles" or "moved in a mixture of many directions." One suggested that the motion might be caused by other factors (e.g., our hand holding it may "shake a little") The last student suggested that the "to-and-fro" vibration of the cup bottom could have triggered the "left-and-right" vibration of the ball. The variation in phenomenal descriptions signaled considerable issues operating the experiment and making stable observations. Although Mr. Pan confirmed the to-and-fro direction and explained away other observations as operational issues, the students had difficulty integrating it into their reasoning. When they got back to the string vibration direction, the first answer was still "a slight up-and-down." The students only conformed to "to-and-fro" after Mr. Pan asked an explicitly guiding question: "So the cup bottom is vibrating like this [gesturing to-and-fro], how would the string vibrate then?"

Mr. Pan then instructed the class to observe how hand-controlled vibration transmits on stretched and curved string phones. The class eventually grasped the idea that the to-and-fro vibration can only transmit through a stretched string. Yet it was the teacher confirmation rather than student MR that dominated the unpack of this black box.

In this instance, the scarcity of reasoning resources constrained the productivity of student MR. In the more productive situations, while empirical evidence might not be accessible, the student could anchor their conjectures on many other resources (e.g., p-prims, prior knowledge, everyday experience) they found relevant, even analogically. In contrast, the question on vibration direction drew a blank when first posed. Students experienced the vibrations on the string as "slight tingle on fingertips" and had no little resource to reason about its direction. This could explain why Mr. Pan's challenge was not taken up as expected. As a bare guess, "multiple direction" can hardly motivate coherence checking. Observations from the later experiment, because of their ambiguity, also failed to serve as solid anchoring resources.

9.6 Discussion

Our analysis illustrated productive and not-so-productive instances of MR from a group of elementary students. Productive MR builds on a rich variety of cognitive resources that can directly or analogically anchor mechanistic conjectures, or resources and manipulations that can afford coherence checking of the mechanistic conjectures. Students can also develop fuller explanations by piecing together the mechanistic details they find relevant. MR is less productive when the goal of establishing canonical explanations overrides the attention to emergent mechanistic discrepancies, or when students do not have the relevant reasoning resources. Now we consider how this work speaks to research and practices on MR.

9.6.1 Understand MR as Processes

It is common for researchers to characterize MR by looking at what components are involved and/or whether cross-scalar thinking is present (Bachtiar et al., 2024; Berg & Hulten, 2024; Dickes et al., 2016). Explicitly or implicitly, such practice treats mechanistic ideas as separate products and assumes an ideal image of quality MR. In ongoing classroom conversations, however, the productivity of MR is a much more complicated issue. For instance, productivity can come from identifying how information from various resources conflict or align with certain proposed mechanistic elements. If further observation of the phenomenon provides information for coherence checking, then phenomenal description, which is often considered less sophisticated in reasoning, can contribute significantly to the explanatory goal. In our case, this occurred when the class screened their initial mechanistic ideas by drawing on the observation of vibration changes along the curved string.

On the other hand, it is not always productive to unpack further mechanistic details. Theoretically, what mechanistic details are relevant depends on what aspect of the phenomenon we want to explain (Craver, 2015; Haskel-Ittah, 2023). Professional scientists can make such decisions based on their research needs. In a classroom setting, it matters *who* –the teacher or the students—perceives the detail as relevant. When students consider certain details as important, especially if they have related discrepancies, overlooking that and rushing to canonical answers could block productive MR and leave students with untreated mechanistic conjectures. Similarly, it could be futile to engage students in unpacking some detail they consider irrelevant, even if it is relevant from an authoritative perspective. An instance in our case was the exploration into the vibration direction of the sound. This spatial organization information is indeed important for understanding why a curved string cannot transmit sound (transmission of longitudinal pressure waves requires tension). Yet the resources students have access to could not afford them to see its relevance or generate reasonable conjectures. If we do want students to engage in MR in such cases, we need to first prepare them with experiences they can understand, so that they can gain resources for further reasoning (e.g., understanding how wave transmits in a slinky may help the students in our case to consider vibration direction).

We see the need to understand MR as part of the inquiry process and focus on how it can keep the effort going in developing more coherent mechanistic explanations (Hammer et al., 2008). This process-oriented view calls for analytical attention to the contextualized functions of student MR in the teaching and learning dynamics.

9.6.2 Potential Principles for Engaging Students in MR

Our initial exploration led to some potential principles for engaging students in MR that can be further examined and revised by future researchers and practitioners.

First, check whether students have the resources to reason something mechanistically. The resources can be theoretical, experiential, and empirical information that one can connect to the subject under reasoning. They do not have to be canonically correct. Yet one cannot reason productively without anchoring their ideas on something they take as true at the moment. We can check the resource availability by posing an initial question and observing students' responses. Explicit reference to multiple anchoring resources and variety in initial accounts marked a potential start, as they set the ground for further MR to come into play. In contrast, when students show difficulty generating explicitly anchored ideas, even if they do make guesses, it is hard to expect further reasoning along the line without introducing or activating new resources.

Second, facilitate the development of resources or manipulations for coherence checking. While students can sometimes come up with their own resources or manipulations, they often need help in data collection and the construction of critical manipulations. Teachers may follow the logic of their reasoning and provide scaffolds in these aspects. It can be difficult to come up with effective guidance to emergent ideas, especially with the limited equipment and material in elementary science classrooms. Interviewing a representative subgroup and preparing some manipulative ideas (e.g., Mr. Shu's "V" shape string) beforehand can be a useful strategy. Separating idea generation and coherence checking into different lessons can also be a way to gain some preparation time.

Third, focus on what students see as relevant. When students have difficulties seeing the relevance of some mechanistic details, it is better to leave such details as black boxes to be unpacked in the future, or, introduce experiences that would afford a sense of relevance. When the mechanistic details students identify as relevant go beyond teacher expectation (e.g., sound and vibration as separate entities), it is still better not to rush toward a canonical account without sufficient attention to these ideas, as they can come back and lead the MR to a standstill. In both we see the importance for students to own their MR, or inquiry practices in general (Engle & Conant, 2002). Productive MR should encourage and facilitate students to explore whether what they see as relevant actually plays a role.

References

Abrams, E., & Southerland, S. (2001). The how's and why's of biological change: How learners neglect physical mechanisms in their search for meaning. *International Journal of Science Education, 23*(12), 1271–1281.

Bachtiar, R. W., Meulenbroeks, R. F., & van Joolingen, W. R. (2024). Understanding how student-constructed stop-motion animations promote mechanistic reasoning: A theoretical framework and empirical evidence. *Journal of Research in Science Teaching, 61*(2), 289–318.

Bechtel, W., & Abrahamsen, A. A. (2013). Thinking dynamically about biological mechanisms: Networks of coupled oscillators. *Foundations of Science, 18*, 707–723.

Bechtel, W., & Richardson, R. C. (1993). *Discovering complexity: Decomposition and localization as strategies in scientific research.* Princeton University Press.

Berg, A., & Hultén, M. (2024). Children's emergent mechanistic reasoning in chemistry: a case study about early primary students' reasoning about the phenomenon of thermal expansion of air. *Chemistry Education Research and Practice, 25*(1), 92–114.

Craver, C. (2007). Constitutive explanatory relevance. *Journal of Philosophical Research, 32*(3), 3–20.

Craver, C. F., & Tabery, J. (2015). Mechanisms in science. *Standford Encyclopedia of Philosophy.* https://plato.stanford.edu/entries/science-mechanisms/

Darden, L. (2006). *Reasoning in biological discoveries: Essays on mechanisms, interfield relations, and anomaly resolution*. Cambridge University Press.

De Andrade, V., Shwartz, Y., Freire, S., & Baptista, M. (2022). Students' mechanistic reasoning in practice: Enabling functions of drawing, gestures and talk. *Science Education, 106*(1), 199–225.

Dickes, A. C., Sengupta, P., Farris, A. V., & Basu, S. (2016). Development of mechanistic reasoning and multilevel explanations of ecology in third grade using agent-based models. *Science Education, 100*(4), 734–776.
DiSessa, A. A. (1993). Toward an epistemology of physics. *Cognition and Instruction, 10*(2–3), 105–225.
Engle, R. A., & Conant, F. R. (2002). Guiding principles for fostering productive disciplinary engagement: Explaining an emergent argument in a community of learners classroom. *Cognition and Instruction, 20*(4), 399–483.
Glennan, S., Illari, P., & Weber, E. (2022). Six theses on mechanisms and mechanistic science. *Journal for General Philosophy of Science*, 1–19.
Haskel-Ittah, M. (2023). Explanatory black boxes and mechanistic reasoning. *Journal of Research in Science Teaching, 60*(4), 915–933.
Haskel-Ittah, M., & Yarden, A. (2017). Toward bridging the mechanistic gap between genes and traits by emphasizing the role of proteins in a computational environment. *Science & Education, 26*, 1143–1160.
Keiner, L., & Graulich, N. (2020). Transitions between representational levels: characterization of organic chemistry students' mechanistic features when reasoning about laboratory work-up procedures. *Chemistry Education Research and Practice, 21*(1), 469–482.
Krist, C., Schwarz, C. V., & Reiser, B. J. (2019). Identifying essential epistemic heuristics for guiding mechanistic reasoning in science learning. *Journal of the Learning Sciences, 28*(2), 160–205.
Machamer, P., Darden, L., & Craver, C. F. (2000). Thinking about mechanisms. *Philosophy of Science, 67*(1), 1–25.
Moreira, P., Marzabal, A., & Talanquer, V. (2019). Using a mechanistic framework to characterise chemistry students' reasoning in written explanations. *Chemistry Education Research and Practice, 20*(1), 120–131.
Russ, R. S., Coffey, J. E., Hammer, D., & Hutchison, P. (2009). Making classroom assessment more accountable to scientific reasoning: A case for attending to mechanistic thinking. *Science Education, 93*(5), 875–891.
Russ, R. S., Scherr, R. E., Hammer, D., & Mikeska, J. (2008). Recognizing mechanistic reasoning in student scientific inquiry: A framework for discourse analysis developed from philosophy of science. *Science Education, 92*(3), 499–525.
Shu, G., Tang, X., & Chen, H. (2023). How does sound travel in a string? *Science and Children, 60*(7), 71–75.
Talanquer, V. (2024). How we simplify, circumvent, or distort causal mechanistic reasoning in chemistry. *Journal of Chemical Education, 101*(5), 1793–1797.
Woodward, J. (2013). II—Mechanistic explanation: Its scope and limits. *Aristotelian Society Supplementary, 87*(1), 39–65. Oxford, UK: Blackwell Publishing Ltd.

Chapter 10
The Nature of Mechanistic Reasoning in Chemistry

Vicente Talanquer

10.1 Introduction

Interest in understanding how to support the development of students' ability to engage in mechanistic reasoning has increased considerably across different science disciplines in recent years (Bachtiar et al., 2022; Dood & Watts, 2022). These studies include analyses of the challenges students face in building mechanistic explanations of natural phenomena (Bolger et al., 2012; Talanquer, 2018a), explorations of the cognitive resources that students deploy when asked to reason mechanistically (de Andrade et al., 2022; Scharlott et al., 2024), characterizations of progression in mechanistic reasoning (Ryan et al., 2023; Scott et al., 2018), and identification of task features that best scaffold the construction of mechanistic explanations and learning to think mechanistically (Caspari & Graulich, 2019; Noyes et al., 2022). Results from these investigations suggest that although mechanistic reasoning shares essential characteristics across fields (Krist et al., 2019; Russ et al., 2008; see also Chap. 4), domain-specific features should be considered when designing, implementing, and assessing instructional tasks in the different disciplines. The central goal of this contribution is thus to characterize distinctive features of mechanistic reasoning in chemistry and discuss how they may affect learning and teaching in this area. Some of these characteristics may not be unique to the discipline but play central or salient roles in the construction of mechanistic explanations of the properties and behaviors of chemical systems. In the following sections, I describe differences in the types of systems analyzed in biology, chemistry, and physics classrooms, discuss what I see as the most distinctive characteristics of mechanistic

---

V. Talanquer (✉)
Department of Chemistry and Biochemistry, University of Arizona, Tucson, AZ, USA
e-mail: vicente@arizona.edu

N. Graulich et al. (eds.), *Exploring Mechanistic Reasoning in Science Education*, Contributions from Science Education Research 17,
https://doi.org/10.1007/978-3-032-10246-1_10

chemical reasoning in chemistry, and then analyze affordances and challenges in teaching and learning mechanistic reasoning in this discipline.

10.2 Differences Across School Disciplines

Although the boundaries between scientific disciplines are diffuse, conventional school science more markedly separates the types of systems that are the subject of attention in traditional biology, chemistry, and physics courses. Conventional curricula and teaching practices in these scientific disciplines thus create distinct opportunities for students to engage in mechanistic reasoning and direct students' attention to different types of systems, components, properties, and interactions, as well as to domain-specific representational and modeling tools (e.g., force diagrams versus chemical formulas).

For example, physics classes typically focus on the properties and behaviors of systems whose behaviors can be explained by analyzing fundamental physical properties, such as the mass or electric charge of a few components. In these cases, natural laws or principles sustain the elaboration of mathematical models of a system (e.g., a sliding object on a ramp or an electric charge near a charged surface) that can generate quantitative predictions of expected behaviors or explain observed phenomena. Building these law-based predictions and explanations is the core activity in most conventional introductory physics courses, where inquiry activities, when present, tend to emphasize pattern-seeking and mathematical modeling versus mechanistic reasoning (Tang et al., 2020).

Conventional chemistry and biology curricula create opportunities for students to elaborate mechanistic accounts to explain observed properties and phenomena in systems comprised of more diverse and a larger number of components. In these environments, domain-specific models are used to build qualitative causal stories to predict or explain the emergence of a property or behavior in a system of interest. In chemistry, for example, these explanations often rely on analyzing physical interactions between relevant model entities (e.g., electrons, atoms, molecules) defined in a reduced set of granularity levels (electronic, atomic, molecular). These interactions determine or alter the relative position, orientation, or structure of system components, affecting their kinetic and potential energy and, consequently, the energetic state of the system.

In biology, a system's relevant model entities are more complex (e.g., cells, organisms, species), defined in a wider range of granularity levels (macromolecular, organelle, cellular, organ, organism, ecosystem), and subject to more diverse interactions (e.g., physical, chemical, ecological) than in conventional chemical systems. This complexity leads to more varied states and the acquisition of apparent functions within the system (van Mil et al., 2013). Given the complexity of the systems under analysis, representations of biological entities or processes are often richer in detail and more descriptive and diagrammatic than in chemistry (Schönborn & Bögeholz, 2013).

Although the types of systems analyzed in biology and chemistry courses are very diverse, basic subcomponents are less varied in chemical than in biological systems. Most systems explored in chemistry classes include components comprised of a small set of different atoms. The relative values of relevant properties of these atoms are well known or can be inferred from their position in the Periodic Table, facilitating the prediction of their behaviors when isolated or combined with other atoms in molecules or polyatomic ions. Additionally, chemists have developed a rich and highly structured representational system that facilitates recognizing and visualizing compositional and structural features that determine the properties and behaviors of the entities of interest and aids in building causal stories about relevant changes (Hoffmann & Laszlo, 1991). The extraordinary qualitative predictive and explanatory power of the representational system in chemistry is unique, as they are the challenges associated with thinking with and about chemical representations (Talanquer, 2022). Learning to extract information from chemical representations of system components is critical for inferring the properties, potential interactions, and behaviors of chemical entities in the systems of interest. This information, in combination with knowledge about fundamental chemical mechanisms (Talanquer, 2018b), is then used to build mechanistic accounts of the targeted phenomena. In the following section, I highlight important particular of this intellectual process. Some of these characteristics may apply to mechanistic reasoning in other disciplines. The goal in this chapter is to highlight their specific manifestation when analyzing and representing chemical systems and discuss their implications for teaching and learning in chemistry.

10.3 Mechanistic Reasoning in Chemistry

As a mode of reasoning, mechanistic reasoning in chemistry has the following characteristics. It is (a) hybrid, (b) semi-encapsulated, and (c) cross-dimensional. In the next subsections, I elaborate on each of these features.

10.3.1 Hybrid Reasoning

Chemistry educators and philosophers suggest that chemical rationales can be classified into model-based, case-based, and rule-based according to the nature of the knowledge used for their construction (Kraft et al., 2010; Talanquer, 2021b). Model-based reasoning uses abstract constructs representing the components of a system (e.g., electrons), their properties (e.g., electric charge), their interactions (e.g., electrostatic forces), and their behaviors (e.g., electron transfer) to build mental or visual representations of substances and processes. When engaged in model-based reasoning, students may generate diverse accounts, noncausal and causal, for the properties and behaviors under analysis (Sevian & Talanquer, 2014). For example, they may rely on relational reasoning by building direct associations between the properties of a component or a system and its behaviors without an explicit justification for the association. Or they may express teleological reasoning, implying that the observed behavior is driven by an entity's explicit or implicit goals (see Chap. 2). They could also build simple-causal rationales connecting properties and behaviors by single causal links or engage in multi-causal reasoning by identifying and relating multiple causes. Mechanistic reasoning is a form of model-based reasoning that relies on building simple-causal and multiple-causal links between properties and behaviors concatenated to generate causal stories (Russ et al., 2008).

While model-based reasoning relies on abstract representations and general theoretical principles, case-based reasoning relies on specific past examples or experiences as a source of knowledge (Kovac, 2002). Over the years, chemists have built extensive sets of classification systems that allow them to condense and organize information about the properties of known substances and reactions (Schummer, 1998). These systems support case-based reasoning in which knowledge about substance or process types allows us to infer their properties and potential behaviors without invoking a model. For example, we can predict that solutions of sodium chloride, NaCl(s), conduct electricity simply because it is a soluble salt. Similarly, we can predict the outcome of the reaction between CH_4(g) and O_2(g) by recognizing the process as the combustion of a hydrocarbon.

When seeking to build mechanistic accounts to make predictions or construct explanations, chemists frequently use case-based reasoning to select the chemical models that will guide their thinking. Imagine, for example, that we wanted to explain why the boiling point of liquid ammonia, NH_3(l), is higher than that of liquid methane, CH_4(l). In this instance, building a mechanistic explanation demands first recognizing that these substances are covalent compounds (case). Thus, they can be modeled as molecular systems, and their relative boiling points can be determined by analyzing the nature and strength of the intermolecular forces between their corresponding molecules. On the other hand, if we must explain the difference in melting points between MgO(s) and NaF(s), cased-based reasoning would direct us to model these two substances as ionic compounds and to evaluate how differences in the electric charge and size of the composing ions may affect their electrostatic interactions. When engaged in mechanistic reasoning in chemistry, case-based reasoning guides the selection of the chemical models to deploy.

Chemical thinking also relies on principles or rules, both empirical and theoretical, that condense knowledge into modeling primitives or reasoning heuristics that simplify sense- and decision-making. Rule-based reasoning allows us to make quick inferences about the nature of relevant entities in a system, their properties, and behaviors. Consider the chemical examples presented in the paragraph above. The symbolic representations of the substances under analysis, NH_3(l), CH_4(l), MgO(s), and NaF(s), convey explicit and implicit information about the chemical nature of their atomic components. The chemical elements N, C, H, O, and F are classified as nonmetallic, while Mg and Na are metallic. Based on this information, we can apply the following empirical rules to infer the nature of the chemical compounds under analysis: Combinations of nonmetallic elements lead to the formation of covalent compounds; combinations of nonmetallic and metallic elements produce ionic compounds. Many rules in chemistry are empirical and not linked to a theoretical model, such as the solubility rules or the Le Chatelier's principle. Others, like the octet rule, can be derived from theoretical models but have strong empirical roots.

When engaged in mechanistic reasoning in chemistry, rule-based reasoning facilitates the generation of inferences about the properties and behaviors of modeled system components. Imagine, for example, that we are interested in building a mechanistic explanation for the dominant production of one of two potential products in the addition reaction of water to an alkene. During this task, it is likely that at some point, we will invoke the theoretical rule "more substituted carbocations are more stable and thus likely to form" without further explanation to justify the outcome.

Mechanistic reasoning in chemistry is often a hybrid form of reasoning that integrates model-based, case-based, and rule-based reasoning when making predictions or building explanations. The back and forth between these different types of reasoning does not only occur when first analyzing a problem and deciding how to model a system. It tends to occur at various moments, particularly when new species form during a chemical process. For example, an intermediate appears in the system, and case-based reasoning is used to categorize it (e.g., tertiary carbocation), rule-based reasoning is used to infer its properties (e.g., tertiary carbocations are more energetically stable), and model-based reasoning is applied to build a causal story about its interactions with other species. During this process, information linked to the chemical representations of the entities under analysis must be unpacked and used to guide sense- and decision-making.

10.3.2 Semi-Encapsulated Reasoning

When building mechanistic explanations in chemistry, we commonly encapsulate detailed information about sub-mechanisms responsible for some properties or events into high-level concepts (Talanquer, 2024). For example, in generating a mechanistic account of a chemical reaction, we may claim that an intermediate carbocation is stabilized by "induction" without specifying how this stabilization occurs. Concepts such as induction, resonance, or hyperconjugation encapsulate electronic mechanisms and are commonly deployed in chemistry to make judgments about the relative energetic stability of chemical species without unpacking. From this perspective, they function as "black boxes" (Haskel-Ittah, 2023), facilitating the generation of mechanistic accounts without needing to express or understand the mechanisms they bundle.

Relying on encapsulation when generating explanations is not unique to chemistry; it is also frequent in other areas, such as biology and medicine (Schmidt & Rikers, 2007). However, in these fields, it is more common not to know all the details of the encapsulated mechanism. In these cases, the high-level concept functions as an actual black box that cannot be fully unpacked (Haskel-Ittah, 2023). In chemistry, the degree to which sub-mechanisms are encapsulated mostly depends on the explanatory goals, the nature of the task, and the context in which the activity takes place. Encapsulation facilitates prediction and explanation by directing our attention to the entities, properties, and behaviors most relevant for accomplishing the task at hand. Encapsulations often pack sub-mechanisms for properties or processes that occur in various chemical systems, functioning as transferable objects that support mechanistic reasoning in diverse contexts.

Encapsulation is pragmatically unavoidable when generating mechanistic accounts in chemistry, as it would be unproductive to unpack every property or event to the lowest level of granularity. Sometimes, encapsulation is layered, and one must decide what level of unpacking is appropriate given the task goals. For example, one may invoke "resonance" to account for the stability of a carboxylate ion, justifying the claim by pointing to the "delocalization" of electrons without unpacking the underlying mechanism in terms of changes in the kinetic and potential energy of electrons. We can expect novice learners not to know what level of encapsulation (or unpacking) is appropriate for a particular task. Conventional rules for when encapsulation is pertinent or expected are not explicit, and it is doubtful that even experts could clearly articulate them.

Systematically encapsulating sub-mechanisms may create knowledge gaps and generate alternative conceptions, particularly when students do not have frequent opportunities to unpack and reflect on the bundled mechanisms. Educational research has shown that, in the absence of normative knowledge, learners often fill "black boxes" with idiosyncratic ideas generated at the moment based on prior knowledge and experiences (Brown, 2014). Student learning benefits when encapsulations are made explicit and learners have opportunities to recognize them in given and self-generated explanations, unpack them, and use them to generate mechanistic accounts in different contexts (Haskel-Ittah, 2023).

10.3.3 Cross-Dimensional Reasoning

Existing research in chemistry education indicates that reasoning in the discipline demands connecting the properties and behaviors of entities modeled at different levels of granularity, from the macroscopic to the electronic scales (Gilbert & Treagust, 2009; Talanquer, 2021b). Applying relevant relationships that connect the properties of the entities at one level with the properties and behavior of the entities at a sub-level or an upper level is a critical skill for building mechanistic explanations in chemistry. Learners should be able to move back and forth between levels as they seek to explain the targeted phenomenon at one scale using models defined at lower scales. For example, in making sense of the relatively high boiling point of liquid water, $H_2O(l)$, learners must zoom in from the macroscopic to the electronic levels and then zoom out, passing through the molecular and multimolecular levels again (see Fig. 10.1). In this process, different models are built, and information

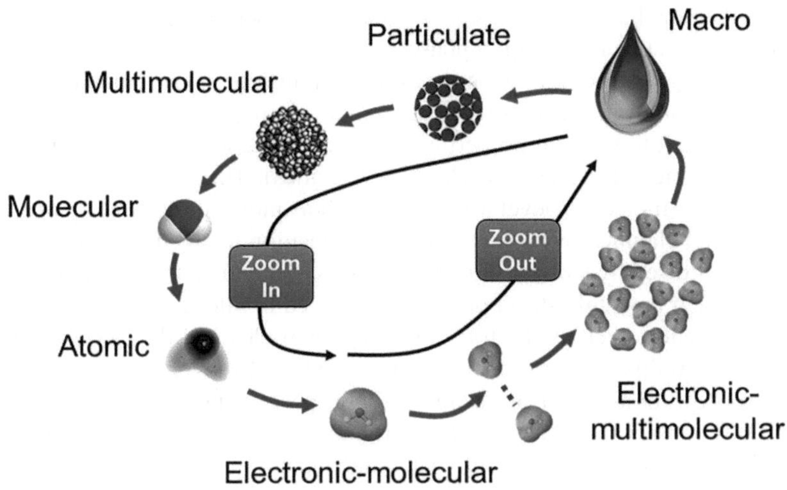

Fig. 10.1 Different levels of granularity explored in building mechanistic explanations in chemistry

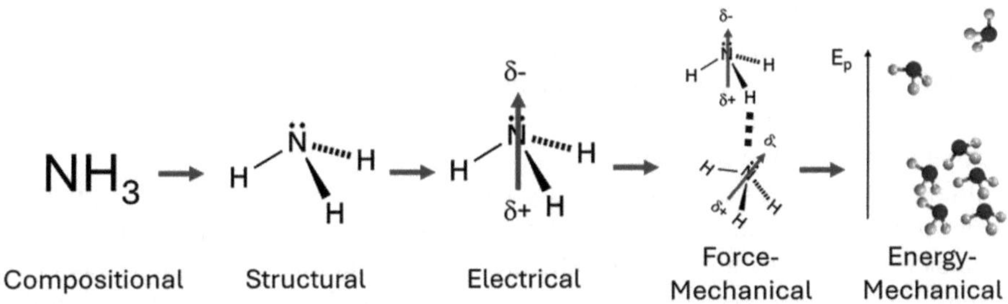

Fig. 10.2 Different physical dimensions of analysis in mechanistic reasoning in chemistry

about the properties, behavior, and organization of the components at one sublevel (e.g., electronic distribution in a water molecule) is used to predict properties of entities at an upper level (e.g., dipole-dipole interactions between polar water molecules) and develop the causal story.

Moving fluidly across granularity levels is only part of the challenge when generating mechanistic accounts in chemistry. This task also demands moving across different physical dimensions of analysis. Consider the case of explaining why the boiling point of $NH_3(l)$ is higher than that of $CH_4(l)$. To build the explanation, one must first move from the compositional to the structural dimensions, as illustrated in Fig. 10.2. Then, from the structural to the electrical dimension to determine charge distribution in the molecule, and from the electrical to a force-mechanical dimension to predict the nature and relative strength of intermolecular interactions. The explanation may also transition from the force-mechanical to the energy-mechanical dimension when claiming that more energy will be required to separate ammonia than methane molecules, given that the former molecules exist in a lower potential energy state in the liquid phase.

Transitioning from one dimension to another in Fig. 10.2 demands analyzing explicit information in the symbolic representations (e.g., types of atoms present) and connecting it to relevant properties of the components (e.g., bonding capacity, number of valence electrons, and electronegativities). This information must then be used to infer molecular-level properties, such as molecular geometry and polarity. This initial chemical modeling phase (composition to electrical dimensions) is frequently followed by a physical modeling phase (electrical to mechanical dimensions), in which molecular-level properties are used to infer relevant aspects of the system mechanics, such as the forces between particles and their effects on their energy states. At this stage, mechanisms are often built by invoking a relevant mechanism prototype, or mechanism schemata, for the process under analysis (Machamer et al., 2000); in the example above, the particulate-level mechanism schemata for a phase change (Talanquer, 2018b). This intellectual process is summarized in Fig. 10.3.

Mechanistic explanations in chemistry frequently involve connecting composition-structure-electron distribution accounts about the properties of relevant chemical entities to mechanical accounts about the nature and strength of their interactions or their energetic states (Caspari et al., 2018). These connections allow us to make predictions about their stability and likely behaviors. The ability to connect information about the chemical identity of the components of a system to their expected physical activities and organization over space and time is a hallmark of mechanistic reasoning in chemistry.

Figure 10.2 makes explicit not only the transitions between different physical dimensions when building mechanistic explanations in chemistry but also across various degrees of iconicity in the chemical representations that support such reasoning. In this context, iconicity is a measure of the degree to which a chemical representation resembles the entity it seeks to depict. For example, when representing a hydrogen atom, the symbol H has low iconicity while a drawing of a small spherical ball has a higher degree of iconicity. Thus, from left to right in Fig. 10.2, the iconicity increases as we move from a symbolic representation of the ammonia

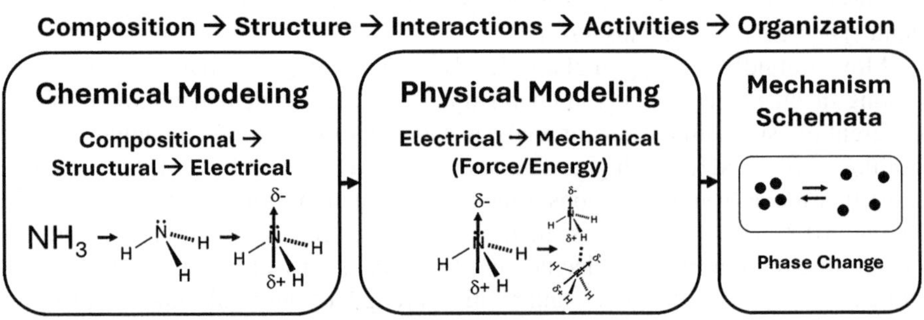

Fig. 10.3 Common stages in the modelling process in mechanistic reasoning in chemistry

molecule, NH_3, to structural representations that depict more literal images of the modeled molecule. Mechanistic accounts of chemical phenomena are often guided, supported, and communicated using representations that vary in their iconicity and explicitness of compositional and structural elements. When building a mechanism, a representation may be unpacked to make some features more explicit and then packed again to direct the attention to other areas. This variability in the depiction of the causal story highlights the high level of representational competency that mechanistic reasoning in chemistry can demand (Talanquer, 2022).

Besides requiring fluent transitions between levels of granularity, different physical dimensions, and degrees of iconicity, mechanistic reasoning in chemistry also demands connecting qualitative and quantitative information about entities in a system. In this regard, this type of reasoning shares many of the characteristics ascribed to "qualitative reasoning" in human cognition and artificial intelligence (Forbus, 2019; Talanquer, 2021a). It relies on both qualitative (e.g., types of atoms, types of reactions) and quantitative (e.g., valence, atomic size, electronegativity) information about the components of a system to generate non-numerical descriptions of their behaviors. Numerical values for properties are not used to make quantitative predictions but rather to enable comparisons and reasoning with inequalities (e.g., larger than, weaker than). For example, decisions about the energetic stability of ions are made based on the analysis of which ion is bigger or has a higher electronegativity rather than referring to precise values. Similarly, predictions about relative reaction rates rely on inferences about whether associated activation energies are lower or higher, independently of the actual numbers.

Using qualitative and quantitative information about the properties of system components to build qualitative causal narratives is a distinctive characteristic of mechanistic reasoning in chemistry. In physics, numerical information is commonly employed to make precise predictions based on quantitative models or laws. In biology, quantitative information for the properties of relevant entities is often not available. The level of complexity of chemical systems is such that quantitative modeling is not practically feasible, but quantitative patterns (e.g., periodic trends) enable qualitative prediction and explanation. Connecting qualitative information to quantitative properties and using them to make comparisons, identify inequalities (e.g., more/less nucleophilic), and infer their consequences for the state or behavior of system components is core to building mechanistic accounts in chemistry.

Figure 10.4 summarizes the four dimensions described in this section across which causal stories in chemistry are built. They include granularity, physical dimension, iconicity, and qual-quant connectivity. Thus, this type of reasoning can be characterized as cross-dimensional, requiring the integration of diverse knowledge types and modes of reasoning and the fluid transition between scales of description, physical dimension of analysis, types of representations, and qualitative and quantitative types of information. Mechanistic reasoning in chemistry is a powerful tool for prediction and explanation but is complex and cognitively demanding. By making explicit its characteristics, we may be in a better position to identify the curricular, instructional, and assessment strategies that support its development.

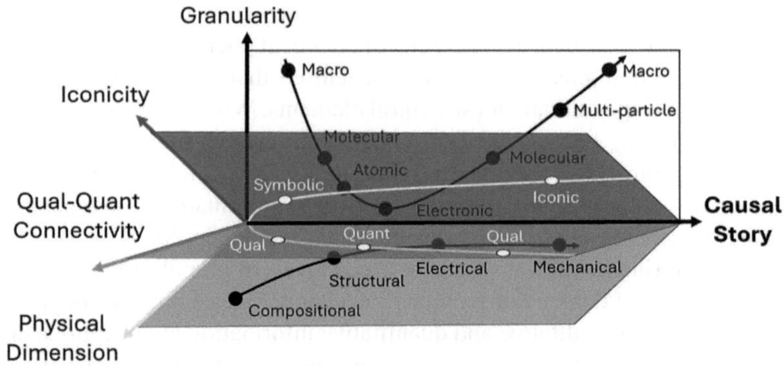

Fig. 10.4 Different dimensions across which mechanistic reasoning in chemistry is built

10.4 Teaching and Learning to Think Mechanistically in Chemistry

Engaging in mechanistic reasoning in chemistry is cognitively taxing as students need to activate different types of knowledge and modes of reasoning to unpack information from chemical representations, categorize entities, infer their properties, and model their behaviors. This intellectual activity often demands a broad and integrated knowledge base and the ability to select and properly apply the types of reasoning that facilitate the construction of the causal story in its various stages. These challenges highlight the importance of carefully identifying and analyzing the fundamental mechanisms we expect students to master and reflecting on the nature of the curriculum, teaching practices, and assessment tools that can better foster student learning. Findings from research in chemistry education in recent years provide insights and guidance in these different areas.

From a curricular perspective, the complex nature of mechanistic reasoning asks for thoughtful sequencing when introducing ideas and articulating learning experiences. Existing research suggests that student reasoning is better advanced when the curriculum follows research-based learning progressions designed to utilize learners' cognitive resources to build and connect new ideas and ways of thinking (Jin et al., 2024). Imagine, for example, that we wanted to help students understand the central chemical idea that differences in the physical and chemical properties of materials stem from variations in their chemical composition. At an early age, we could begin by engaging young children in analyzing objects of varying sizes and shapes made from the same or different materials. Students could be asked to examine and describe these objects, compare their perceptual properties, and categorize them into groups. Next, they might generate simple models to explain a key distinguishing property for each category (e.g., what causes a material to be more, or less, rigid or flexible). The sequence from categorization to modeling tasks makes explicit a productive way of combining case-based and model-based reasoning in chemistry to build mechanistic explanations.

Further exploration could involve manipulating the materials by cutting or grinding them to compare whole-part properties. This activity could scaffold the conceptual differentiation between extensive and intensive properties, helping learners recognize that intensive properties are better suited for distinguishing materials (Wiser et al., 2013). Students could then be guided to investigate various types of intensive properties, starting with mechanical properties and progressing to thermal, electrical, and chemical properties (Ngai et al., 2014). As students explore materials by testing their properties, they could engage in modeling activities to explain observed behaviors. Research suggests that, with appropriate support, elementary school students can successfully develop particulate models of matter to explain changes in materials caused by mechanical action (Acher et al., 2007). These models would have to be revised and refined to align with experimental data from analyzing other properties, from thermal to electrical to chemical in later educational stages. This progression opens opportunities for students to build and compare models, gradually incorporating components at different levels of granularity and properties across diverse physical dimensions.

Throughout this learning sequence, students should be encouraged to create various visual representations to articulate their thinking and communicate their ideas effectively (Cooper et al., 2017). Analyzing and discussing these representations can help students recognize how diverse symbols and icons can convey different information and develop representational competence (ability to interpret, build, and translate between representations with different degrees of iconicity) (Kozma & Russell, 2005). Introducing the representational systems used in chemistry can thus be gradual and connected to representations initially generated by students. Ultimately, we must intentionally create opportunities for students to develop fluency in interpreting, generating, and translating between the diverse chemical representations used to communicate information about the structure, properties, and behaviors of core components in representative systems and processes (Galloway et al., 2017; Webber & Flynn, 2018).

Overall, this example illustrates how learning can be sequenced to create opportunities for students to recognize the various ways of reasoning, levels of granularity, physical dimensions, and degrees of iconicity used in building mechanistic explanations about the properties and behaviors of chemical systems. Chemistry curricula at all educational levels must be designed to ensure that these types of opportunities are not haphazard or sporadic but well-planned, frequent, and coherent. Current educational standards for K-12 in various countries, including the US (National Research Council, 2013), provide frameworks that support this goal. Examples of college-level chemistry curricula that emphasize the development of mechanistic reasoning are also available (Cooper & Klymkowsky, 2013; Cooper et al., 2019; Talanquer & Pollard, 2010).

Thinking mechanistically in chemistry is also fostered by anchoring instructional tasks on the analysis of relevant phenomena (Ralph et al., 2022; Scharlott et al., 2024; see Chap. 12). Some of these tasks should create opportunities for students to make sense of data, express initial ideas based on prior knowledge and experiences, engage in chemical and physical modeling of the systems of interest, use a variety of representations to communicate ideas, generate mechanistic explanations, receive feedback, evaluate and reflect on their ideas and ways of reasoning, and revise their models (Windschitl et al., 2008). Other instructional activities may engage students in the analysis of contrasting models, representations, or ways of reasoning about phenomena to make explicit differences and similarities, scope and limitations, or range of application. Developing mechanistic reasoning requires time and multiple and diverse opportunities for students to learn how to productively transition between the different levels of granularity, physical dimension, iconicity, and qual-quant connectivity that the analysis of chemical tasks demands.

Research in chemistry education highlights specific types of tasks that support the development of mechanistic reasoning within the discipline. Students' ability to connect different levels of granularity, physical dimensions, degrees of iconicity, and qualitative and quantitative information can be enhanced by tasks that differ on the types of compositional and structural representational features (explicit or implicit) that must be unpacked to extract relevant information about the properties of a system's components, or that require learners to pack such information using different types of chemical representations when generating their own models (Graulich et al., 2019; Talanquer, 2022). For example, students engaged in analyzing a model to explain how and why UV radiation leads to the destruction of ozone in the stratosphere may be tasked with translating qualitative compositional/structural information in a reaction mechanism into energetic information represented in an energy diagram and using this qualitative diagram to make quantitative predictions about the effect of various factors on reaction rate (time dimension). In these activities, students will likely require explicit guidance and support learning to connect explicit representational features (e.g., chemical formulas of reactants and products) and implicit properties (e.g., the relative potential energy of reactants and products).

Tasks that require comparing and contrasting chemical entities or processes also foster mechanistic reasoning by helping students discriminate between relevant and irrelevant factors affecting a system's properties and behaviors in different contexts (Graulich & Schween, 2018; see also Chap. 3). These tasks may involve comparing explicit and implicit features in the representations of different substances or reactions or making and justifying predictions about their expected properties and behaviors. In these situations, student learning seems to benefit from using explicit guides that facilitate connecting properties and behaviors at different levels of granularity and translating information between different physical dimensions of analysis (e.g., structural to energy dimensions) (Caspari & Graulich, 2019).

Considering the common stages in the modeling process in mechanistic reasoning in chemistry (see Fig. 10.3), students' ability to think mechanistically also benefits from making explicit the fundamental mechanisms (mechanism schemata) commonly used in chemistry to make predictions or build explanations in different areas (Flynn & Ogilvie, 2015; Talanquer, 2018b). This can be accomplished by designing instructional tasks that require students to reflectively engage in selecting or generating such prototypical mechanisms in various relevant contexts (Cooper et al., 2016). For example, building models that involve different components and interactions but rely on the same mechanism schemata, like models explaining how and why thermal energy is generated when dissolving a chemical substance in water or inducing a combustion reaction.

Mechanistic reasoning in chemistry can also be fostered by openly modeling how we think and promoting reflection on the different types of reasoning we deploy (i.e., case-based, model-based, rule-based) and when and how we do it based on task goals. We should also discuss when and how we use encapsulation to simplify a mechanistic explanation and create spaces for unpacking the encapsulated knowledge to ensure students can independently open these black boxes when necessary (Talanquer, 2024). We often assume that students will tacitly learn the "rules of the game," but in most cases, they finish our courses without appropriating the discourse practices we seek them to embrace (Moon et al., 2016).

From the assessment perspective, educational research provides insights into how to make students' mechanistic reasoning more visible to facilitate providing formative feedback. Task prompts should, for example, explicitly require students to discuss how a process occurs and why it happens (Cooper et al., 2016; Noyes et al., 2022; see Chap. 6), generate explanations for contrasting behaviors based on information unpacked from chemical representations (Kararo et al., 2019), or build visual sketches of mechanistic accounts of target properties and phenomena (Cooper et al., 2017). Traditional summative assessments in chemistry often use close-ended questions that target fragmented pieces of knowledge (e.g., infer properties, predict a behavior, identify mechanistic steps). Mechanistic reasoning can be better assessed using three-dimensional tasks built using performance expectations that reflect the ability to integrate the central ideas, core science and engineering practices, and crosscutting ways of reasoning that students should master (Laverty et al., 2016; Stowe & Cooper, 2019; Talanquer, 2019).

10.5 Final Comments

If we share the belief that building mechanistic accounts of relevant phenomena fosters meaningful learning in a discipline, then we must recognize the benefits to educators and educational researchers from developing a better understanding of the intellectual mechanisms supporting the construction of mechanistic explanations in that domain. Making explicit the required domain-specific knowledge and ways of reasoning provides insights into the educational practices that can help students overcome the cognitive challenges they may face.

Given the complexity of mechanistic reasoning in chemistry, one may be tempted to suggest that only students in advanced courses should be expected to engage in this mode of thinking. Largely, conventional chemistry curricula are built under such an assumption, frontloading the accumulation of vast amounts of declarative knowledge in introductory courses and assuming that students will be able to transfer and productively apply such knowledge when asked to reason with it in upper-division classes or their graduate studies. Education research has demonstrated the many flaws of this approach, which, among other things, tends to promote shallow and fleeting learning, demotivates students, and ignores the challenges of knowledge transfer (National Research Council, 2000).

Many studies have demonstrated that students from very young ages can think about complex systems and reason mechanistically when supported by strong scaffolds (Acher et al., 2007; Wiser et al., 2013; Yoon et al., 2018). Learners can develop a solid knowledge base while acquiring mechanistic reasoning skills when analyzing phenomena of interest to them. Several of the investigations cited in this contribution point to the types of curricular frameworks, instructional practices, and assessment strategies that foster such integration. The major challenge is transforming our educational systems to allow these approaches to flourish.

But even within traditional structures, individual instructors can create more opportunities for students to engage in mechanistic reasoning. Research has shown that individual educators can have a major impact on the type of reasoning that manifests in classrooms through the tasks they ask students to complete and the explanations they provide (Moreira et al., 2019; Reid et al., 2022; Scharlott et al., 2024; Talanquer, 2024). However, instructors must understand the various cognitive challenges that students are likely to encounter and reflect on how their decisions and actions affect the development of student reasoning abilities.

References

Acher, A., Arcà, M., & Sanmartí, N. (2007). Modeling as a teaching learning process for understanding materials: A case study in primary education. *Science Education, 91*(3), 398–418.

Bachtiar, R. W., Meulenbroeks, R. F. G., & van Joolingen, W. R. (2022). Mechanistic reasoning in science education: A literature review. *Eurasia Journal of Mathematics, Science and Technology Education, 18*(11), em2178.

Bolger, M. S., Kobiela, M., Weinberg, P. J., & Lehrer, R. (2012). Children's mechanistic reasoning. *Cognition and Instruction, 30*(2), 170–206.

Brown, D. E. (2014). Students' conceptions as dynamically emergent structures. *Science & Education, 23*(7), 1463–1483.

Caspari, I., & Graulich, N. (2019). Scaffolding the structure of organic chemistry students' multivariate comparative mechanistic reasoning. *International Journal of Physics and Chemistry Education, 11*, 31–43.

Caspari, I., Kranz, D., & Graulich, N. (2018). Resolving the complexity of organic chemistry students' reasoning through the lens of a mechanistic framework. *Chemistry Education Research and Practice, 19*(4), 1117–1141.

Cooper, M., & Klymkowsky, M. (2013). Chemistry, life, the universe, and everything: A new approach to general chemistry, and a model for curriculum reform. *Journal of Chemical Education, 90*(9), 1116–1122.

Cooper, M. M., Kouyoumdjian, H., & Underwood, S. M. (2016). Investigating students' reasoning about acid–base reactions. *Journal of Chemical Education, 93*(10), 1703–1712.

Cooper, M. M., Stieff, M., & DeSutter, D. (2017). Sketching the invisible to predict the visible: From drawing to modeling in chemistry. *Topics in Cognitive Science, 9*(4), 902–920.

Cooper, M. M., Stowe, R. L., Crandell, O. M., & Klymkowsky, M. W. (2019). Organic chemistry, life, the universe and everything (oclue): A transformed organic chemistry curriculum. *Journal of Chemical Education, 96*(9), 1858–1872.

de Andrade, V., Shwartz, Y., Freire, S., & Baptista, M. (2022). Students' mechanistic reasoning in practice: Enabling functions of drawing, gestures and talk. *Science Education, 106*(1), 199–225.

Dood, A. J., & Watts, F. M. (2022). Mechanistic reasoning in organic chemistry: A scoping review of how students describe and explain mechanisms in the chemistry education research literature. *Journal of Chemical Education, 99*(8), 2864–2876.

Flynn, A. B., & Ogilvie, W. W. (2015). Mechanisms before reactions: A mechanistic approach to the organic chemistry curriculum based on patterns of electron flow. *Journal of Chemical Education, 92*(5), 803–810.

Forbus, K. D. (2019). *Qualitative representations*. MIT Press.

Gilbert, J., & Treagust, D. F. (2009). *Multiple representations in chemical education*. Springer.

Graulich, N., Hedtrich, S., & Harzenetter, R. (2019). Explicit versus implicit similarity – Exploring relational conceptual understanding in organic chemistry. *Chemistry Education Research and Practice, 20*(4), 924–936.

Graulich, N., & Schween, M. (2018). Concept-oriented task design: Making purposeful case comparisons in organic chemistry. *Journal of Chemical Education, 95*(3), 376–383.

Haskel-Ittah, M. (2023). Explanatory black boxes and mechanistic reasoning. *Journal of Research in Science Teaching, 60*(4), 915–933.

Hoffmann, R., & Laszlo, P. (1991). Representation in chemistry. *Angewandte Chemie International Edition in English, 30*(1), 1–16.

Jin, H., Yan, D., & Krajcik, J. (2024). *Handbook of research on science learning progressions* (1st ed.). Routledge.

Kararo, A. T., Colvin, R. A., Cooper, M. M., & Underwood, S. M. (2019). Predictions and constructing explanations: An investigation into introductory chemistry students' understanding of structure–property relationships. *Chemistry Education Research and Practice, 20*(1), 316–328. https://doi.org/10.1039/C8RP00195B

Kovac, J. (2002). Theoretical and practical reasoning in chemistry. *Foundations of Chemistry, 4*(2), 163–171.

Kozma, R., & Russell, J. (2005). Students becoming chemists: Developing representational competence. In J. K. Gilbert (Ed.), *Visualization in science education* (pp. 121–145). Springer.

Kraft, A., Strickland, A. M., & Bhattacharyya, G. (2010). Reasonable reasoning: Multi-variate problem-solving in organic chemistry. *Chemistry Education Research and Practice, 11*(4), 281–292.

Krist, C., Schwarz, C. V., & Reiser, B. J. (2019). Identifying essential epistemic heuristics for guiding mechanistic reasoning in science learning. *Journal of the Learning Sciences, 28*(2), 160–205.

Laverty, J. T., Underwood, S. M., Matz, R. L., Posey, L. A., Carmen, J. H., Caballero, M. D., Fata-Hartley, C. L., Jardeleza, S. E., & Cooper, M. M. (2016). Characterizing college science assessments: The three-dimensional learning assessment protocol. *PLoS One, 11*(9), e0162333.

Machamer, P., Darden, L., & Craver, C. F. (2000). Thinking about mechanisms. *Philosophy of Science, 67*(1), 1–25.

Moon, A., Stanford, C., Cole, R., & Towns, M. (2016). The nature of students' chemical reasoning employed in scientific argumentation in physical chemistry. *Chemistry Education Research and Practice, 17*(2), 353–364. https://doi.org/10.1039/C5RP00207A

Moreira, P., Marzabal, A., & Talanquer, V. (2019). Investigating the effect of teacher mediation on student expressed reasoning. *Chemistry Education Research and Practice, 20*(3), 606–617. https://doi.org/10.1039/C9RP00075E

National Research Council. (2000). *How people learn: Brain, mind, experience, and school: Expanded edition*. The National Academies Press.

National Research Council. (2013). *Next generation science standards*. The National Academies Press.

Ngai, C., Sevian, H., & Talanquer, V. (2014). What is this substance? What makes it different? Mapping progression in students' assumptions about chemical identity. *International Journal of Science Education, 36*(14), 2438–2461.

Noyes, K., Carlson, C. G., Stoltzfus, J. R., Schwarz, C. V., Long, T. M., & Cooper, M. M. (2022). A deep look into designing a task and coding scheme through the lens of causal mechanistic reasoning. *Journal of Chemical Education, 99*(2), 874–885.

Ralph, V. R., Scharlott, L. J., Schwarz, C. E., Becker, N. M., & Stowe, R. L. (2022). Beyond instructional practices: Characterizing learning environments that support students in explaining chemical phenomena. *Journal of Research in Science Teaching, 59*(5), 841–875.

Reid, J. W., Kirbulut Gunes, Z. D., Fateh, S., Fatima, A., Macrie-Shuck, M., Nennig, H. T., Quintanilla, F., States, N. E., Syed, A., Cole, R., Rushton, G. T., Shah, L., & Talanquer, V. (2022). Investigating patterns of student engagement during collaborative activities in undergraduate chemistry courses. *Chemistry Education Research and Practice, 23*(1), 173–188. https://doi.org/10.1039/D1RP00227A

Russ, R. S., Scherr, R. E., Hammer, D., & Mikeska, J. (2008). Recognizing mechanistic reasoning in student scientific inquiry: A framework for discourse analysis developed from philosophy of science. *Science Education, 92*(3), 499–525.

Ryan, Z., Danish, J., Zhou, J., Stiso, C., Murphy, D., Duncan, R., Chinn, C., & Hmelo-Silver, C. E. (2023). Investigating students' development of mechanistic reasoning in modeling complex aquatic ecosystems. *Frontiers in Education, 8*, 1159558.

Scharlott, L. J., Rippey, D. W., Rosa, V., & Becker, N. M. (2024). Progression toward causal mechanistic reasoning through phenomenon-based learning in introductory chemistry. *Journal of Chemical Education., 101*, 777–788.

Schmidt, H. G., & Rikers, R. M. (2007). How expertise develops in medicine: Knowledge encapsulation and illness script formation. *Medical Education, 41*(12), 1133–1139.

Schönborn, K. J., & Bögeholz, S. (2013). Experts' views on translation across multiple external representations in acquiring biological knowledge about ecology, genetics and evolution. In D. F. Treagust & C.-Y. Tsui (Eds.), *Multiple representations in biological education* (pp. 111–128). Springer.

Schummer, J. (1998). The chemical core of chemistry i: A conceptual approach. *Hyle: International Journal for Philosophy of Chemistry, 4*, 129–162.

Scott, E. E., Anderson, C. W., Mashood, K. K., Matz, R. L., Underwood, S. M., & Sawtelle, V. (2018). Developing an analytical framework to characterize student reasoning about complex processes. *CBE Life Sciences Education, 17*(3), ar49.

Sevian, H., & Talanquer, V. (2014). Rethinking chemistry: A learning progression on chemical thinking. *Chemistry Education Research and Practice, 15*(1), 10–23.

Stowe, R. L., & Cooper, M. M. (2019). Assessment in chemistry education. *Israel Journal of Chemistry, 59*(6–7), 598–607.

Talanquer, V. (2018a). Exploring mechanistic reasoning in chemistry. In J. Yeo, T. W. Teo, & K.-S. Tang (Eds.), *Science education research and practice in Asia-pacific and beyond* (pp. 39–52). Springer.

Talanquer, V. (2018b). Importance of understanding fundamental chemical mechanisms. *Journal of Chemical Education, 95*(11), 1905–1911.

Talanquer, V. (2019). Some insights into assessing chemical systems thinking. *Journal of Chemical Education, 96*(12), 2918–2925.

Talanquer, V. (2021a). Chapter 2. Qualitative reasoning in problem-solving in chemistry. In G. Tsaparlis (Ed.), *Problems and problem solving in chemistry education* (pp. 15–37). The Royal Society of Chemistry.

Talanquer, V. (2021b). Multifaceted chemical thinking: A core competence. *Journal of Chemical Education, 98*(11), 3450–3456.

Talanquer, V. (2022). The complexity of reasoning about and with chemical representations. *JACS Au, 2*(12), 2658–2669.

Talanquer, V. (2024). How we simplify, circumvent, or distort causal mechanistic reasoning in chemistry. *Journal of Chemical Education, 101*(5), 1793–1797.

Talanquer, V., & Pollard, J. (2010). Let's teach how we think instead of what we know. *Chemistry Education Research and Practice, 11*(2), 74–83. https://doi.org/10.1039/C005349J

Tang, X., Elby, A., & Hammer, D. (2020). The tension between pattern-seeking and mechanistic reasoning in explanation construction: A case from Chinese elementary science classroom. *Science Education, 104*(6), 1071–1099.

van Mil, M. H. W., Boerwinkel, D. J., & Waarlo, A. J. (2013). Modelling molecular mechanisms: A framework of scientific reasoning to construct molecular-level explanations for cellular behaviour. *Science & Education, 22*(1), 93–118.

Windschitl, M., Thompson, J., & Braaten, M. (2008). Beyond the scientific method: Model-based inquiry as a new paradigm of preference for school science investigations. *Science Education, 92*(5), 941–967.

Wiser, M., Frazier, K. E., & Fox, V. (2013). At the beginning was amount of material: A learning progression for matter for early elementary grades. In G. Tsaparlis & H. Sevian (Eds.), *Concepts of matter in science education* (pp. 95–122). Springer.

Yoon, S. A., Goh, S.-E., & Park, M. (2018). Teaching and learning about complex systems in k–12 science education: A review of empirical studies 1995–2015. *Review of Educational Research, 88*(2), 285–325.

Chapter 11
Modality Matters: Exploring and Capturing Students' Mechanistic Reasoning During Computational Modeling in Science Education

Tamar Fuhrmann, Leah F. Rosenbaum, Aditi Wagh, Adelmo Eloy, Paulo Blikstein, and Michelle H. Wilkerson

11.1 Introduction

There is a growing consensus that mechanistic reasoning (MR) is a crucial form of reasoning in science education (Krist et al., 2019; Machamer et al., 2000; Pluta et al., 2011; Russ et al., 2008; Salmon, 1979). MR involves the identification of factors and relationships that underlie a phenomenon (Krist et al., 2019; Louca et al., 2011; Machamer et al., 2000; Perkins & Grotzer, 2000; Russ et al., 2008; Springer & Keil, 1991). Students often struggle to translate their understanding of a phenomenon into MR elements in part because they may not recognize the underlying relationships between elements of a system (Hmelo-Silver & Azevedo, 2006; Hmelo-Silver & Pfeffer, 2004; Russ et al., 2008; Wilensky & Resnick, 1999). Thus it is critical that researchers and practitioners understand what helps students develop MR, the best methods to surface these skills, and how to support and enhance their understanding of scientific phenomena.

One way to support students in developing mechanistic explanations is to engage them in creating and refining agent-based models (ABM). Much of the work around ABM and mechanistic reasoning has shown that ABM supports engaging in mechanistic

T. Fuhrmann (✉) · L. F. Rosenbaum · A. Eloy · P. Blikstein
Transformative Learning Technologies Laboratory, Department of Mathematics, Science & Technology, Teachers College, Columbia University, New York, NY, USA
e-mail: tf2464@tc.columbia.edu

A. Wagh
MIT Scheller Teacher Education Program, Cambridge, MA, USA

The Concord Consortium, Concord, MA, USA

M. H. Wilkerson
University of California Berkeley School of Education, Berkeley, CA, USA

N. Graulich et al. (eds.), *Exploring Mechanistic Reasoning in Science Education*, Contributions from Science Education Research 17,
https://doi.org/10.1007/978-3-032-10246-1_11

reasoning as learners encode properties and behaviors for individual agents and observe the aggregate outcomes from their interactions (Dickes et al., 2016; Eberbach et al., 2021; Louca & Zacharia, 2007; Wilensky, 2003; Wilkerson et al., 2015b; Xiang & Passmore, 2015). However, this special alignment between the representational features of ABM and students' mechanistic reasoning can present a methodological challenge. This is because modalities like drawings or written text lack the representational specificity of ABM to capture details of mechanistic reasoning. Therefore, it is important to investigate how students express their mechanistic reasoning across other modalities used in studies and curricula to determine what aspects of their comprehension may be captured or missed. This understanding can aid researchers in more effectively studying students' developing understandings over time, as well as curriculum designers and educators in capturing and responding to students' thinking.

In this chapter, we explore the relationships between the non-computational modes of expression (utterances, drawings, and written responses) students use to describe their work within a computational modeling unit and the elements of MR students demonstrate in their explanations of the target phenomenon. This study builds on research about 6th-grade students' developing understanding of diffusion within a unit in *MoDa* (https://moda.education/), an agent-based computational modeling environment. Using Russ et al.'s mechanistic reasoning framework (2008), we study how learners expressed five elements of MR across a week-long unit. We document when and how these five elements differentially appear across three expressive modalities that are typically used in ABM studies and curricula: written pre-/post-unit tests, drawings, and verbal presentations. Our findings suggest that not only do the different elements of MR develop at different rates, but also that learners express each element of MR to different degrees across differing modalities. The implications for teaching, assessment, and research include a call for sensitivity to how non-computational modalities differentially afford the documentation and expression of MR.

11.2 Theoretical Background

In this section, we review existing research on the relationship between MR and ABM, including how activities such as drawing, discussion, and explanation support the development and understanding of MR. Whereas these activities are often a part of curricular activities, less work has focused on their specific role in capturing students' expressions of MR for research or assessment purposes. Our contributions focus on the various representational media through which MR emerges.

11.2.1 Mechanistic Reasoning and Agent-Based Models

Prior work has investigated the benefits of agent-based modeling to support mechanistic reasoning (e.g., Dickes et al., 2016; Hsiao et al., 2019; Wilkerson et al., 2017; Blikstein & Wilensky, 2009; Wagh & Wilensky, 2018; Wilkerson et al., 2015b).

Students develop mechanistic explanations as they build agent-based models (Wilkerson et al., 2017), which require them to specify how individual components and their interactions give rise to complex systems and phenomena. As ABM involves defining the rules and behaviors of individual agents, it prompts students to consider the mechanisms driving these interactions (Machamer et al., 2000). Creating and experimenting with these models also enables students to test hypotheses and theories by running simulations to explore different ideas.

Observing how students describe and explain mechanisms while constructing a scientific model can shed light on the development of their mechanistic reasoning. For instance, Russ et al. (2008) observed that shifts in mechanistic reasoning often occur when students transition from describing phenomena in general in a "show-and-tell" manner to identifying the entities, activities, and properties of complex systems during interaction (p. 520). Interestingly, the shift often follows particular prompts from a teacher (e.g., "Why do you think that is? Why did that happen? How could it be that we all got these different results?") and persists for several minutes before students then transition back to lower-level, more generally descriptive language. The authors attribute this phenomenon to a developmental process whereby introductory levels of mechanistic reasoning, such as identifying the entities and behaviors in a system, might serve as "building blocks" to deeper forms of reasoning, such as describing the organization of entities or the causal chains that connect them (p. 521).

Over the past two decades, advancements in agent-based, domain-specific, block-based computational modeling technologies have enabled learners to more easily create models using visual and block-based programming in environments such as AgentSheets (Repenning, 2017), NetTango (Horn et al., 2014), Deltatick (Wilkerson et al., 2015b), ViMap (Sengupta et al., 2021), StarLogo Nova (MIT Scheller Teacher Education Program, 2019) and Much.Matter.in.Motion (Saba et al., 2021). MoDa, the computational modeling platform used in this study (Fuhrmann et al., 2022; Wagh et al., 2022), belongs to this genre. It builds on prior work with block-based and domain-specific approaches while adding support for comparing models to data and for multiple levels of model specificity through "unpacking" blocks (Eloy et al., n.d. under review).

11.2.2 The Use of Other Media to Support and Capture MR

Alongside these technological developments, a variety of supportive activities such as planning, sketching, animation, and other forms of modeling in complementary media have been shown to further support students' engagement in MR with ABM (Levy, 2013; Louca & Zacharia, 2007). Embodied activities have helped even very young students express and encode the behaviors of entities in a complex system (Danish et al., 2020; Langbeheim & Levy, 2018). Wilkerson et al. (2015a) leveraged drawing and stop-motion animation to emphasize the importance of well-defined behaviors and interactions to describe evaporation and condensation before students modeled these phenomena using ABM technologies. Similarly, Bachtiar et al. (2024; Chap. 8 in this book) established a framework for how constructing stop-motion animations contributes to students' MR in physics.

While the research described above focuses on the pedagogical role of different modalities in supporting mechanistic reasoning (MR), other research highlights how they shape students' demonstration and assessment of MR. In one study exploring mechanistic reasoning across a variety of university courses, students were asked to choose from a drawing or verbal explanation or to provide both a drawing and a verbal explanation to address various mechanism-related questions (Shiroda et al., 2024). Berg & Hultén (2023) documented how young children learned and expressed thermodynamic understandings across multiple modalities, including verbal expression, dance, and drawing. They described how dance highlighted micro processes but restricted the representation of random three-dimensional motion while drawing foregrounded system organization. More specific to our current focus on MR through ABM, the Connected Chemistry project, which documented students' developing mechanistic understandings as they interacted with agent-based visualizations of basic chemical processes, found that students demonstrated more knowledge and focused on more dynamic elements of chemical transformations when drawing versus when using symbolic or verbal expression (Ryan & Stieff, 2019).

11.3 Methods and Materials

11.3.1 MoDa: Integrating Agent-Based Modeling and Visualizations of Real-World Data

Building on prior research that developed and evaluated domain-specific computational modeling platforms (e.g., Wilkerson et al., 2015b), MoDa is a web-based environment that combines creating computational models using domain-specific code blocks with visualizations of real-world data (Eloy et al., 2022; Fuhrmann et al., 2022; Wagh et al., 2022). MoDa enacts a core idea of the Bifocal Modeling framework (Blikstein, 2014) by embedding data visualizations within the computational modeling environment alongside ways of exploring the phenomenon to support learners in testing their ideas by comparing simulations of those ideas against real-world data.

In MoDa, students build models using domain-specific code blocks (Wilkerson et al., 2015b) to express a variety of common ideas about phenomena within that domain (Fuhrmann et al., 2024). For example, in modeling the diffusion of ink in water, students could use a canonical "bounce" block or a non-canonical "attach" or "erase" block to model particle interactions. Students then refine and validate their models by comparing their computer modeling results with video or quantitative data reflecting the real phenomenon. For example, when modeling the non-canonical "attach" particle interaction, which was drawn from previous work on students' extended studies of diffusion (Fuhrmann et al., 2022), students often notice that the simulated molecules form clumps that contrast with what they see in the video data, leading them away from this non-canonical theory of diffusion.

MoDa consists of (A) a modeling environment and (B) a real-world data area (Fig. 11.1). The modeling environment includes (1) a coding area where students can drag and drop domain-specific blocks (based on Google's Blockly library); (2) a simulation area (built on top of the NetLogo engine, Wilensky, 1999) in which students can see the result of their code and control different variables in the simulation, such as temperature; and (3) data visualizations that illustrate the modeling results in graphs. Figure 11.2 details the available domain-specific blocks—organized into categories of general properties, actions, and control mechanisms—available for programming diffusion models.

11.3.2 Participants

The study took place in a school in California with students from two 6th-grade science classes. White students comprise 74% of the student population, while two or more races account for 18%, Asian or Pacific Islander students represent 5%, Hispanic students constitute 2%, and Black students make up 1%. The student-teacher ratio is 5:1. Across the two classes, 16 students consented to participate in this study (eight girls, six boys, and two non-binary students). The science teacher was part of this project for 2 years and participated in professional development and co-design sessions with the project team.

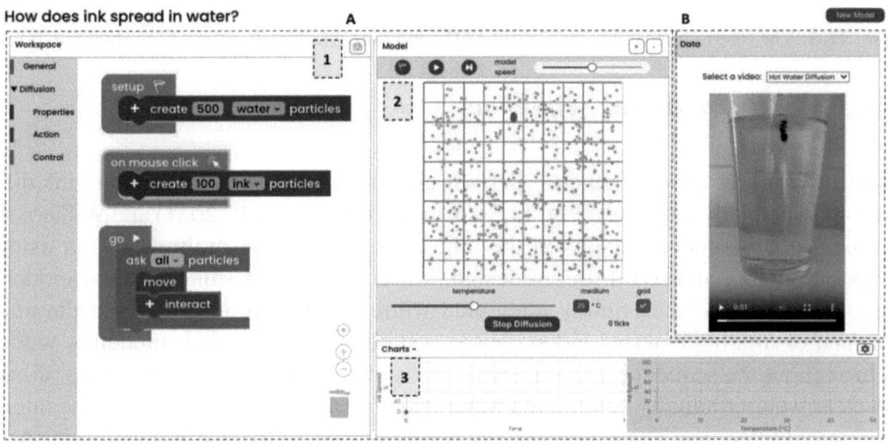

Fig. 11.1 The MoDa modeling and data environment for ink diffusion in water consists of a modeling environment (**a**) and a real-world data area (**b**). The modeling environment includes (1) a coding area, (2) a simulation area, and (3) data visualizations

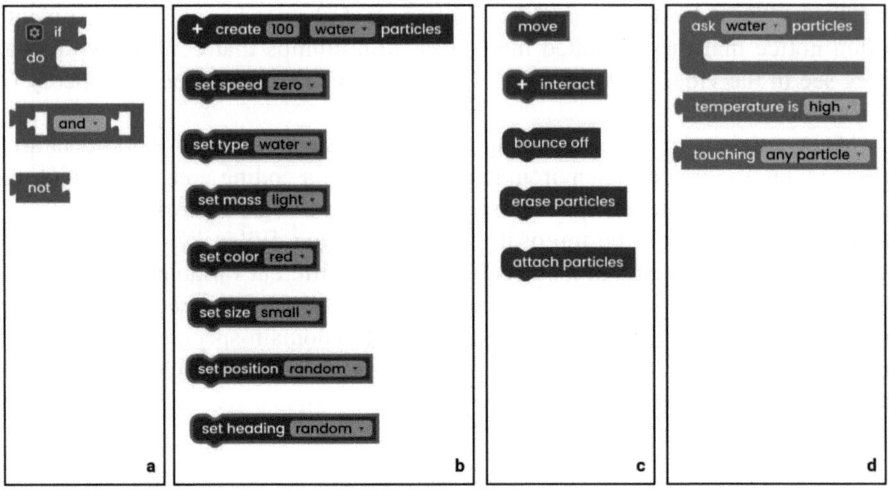

Fig. 11.2 MoDa's available domain-specific blocks are organized into categories: (**a**) general, (**b**) properties, (**c**) actions, and (**d**) control mechanisms

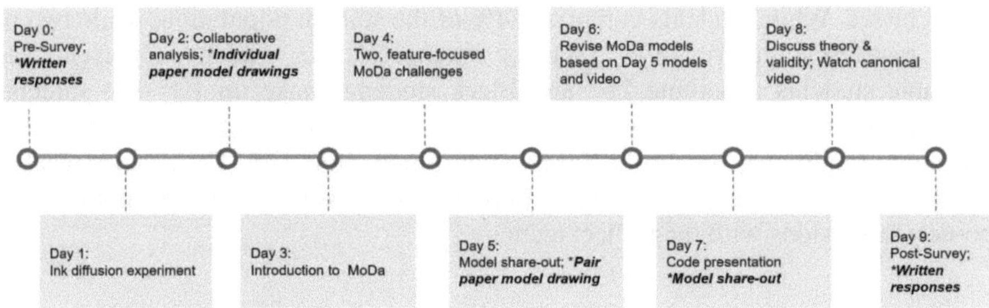

Fig. 11.3 The instructional sequence of the Diffusion unit. Study data sources are shown in bold

11.3.3 Instructional Sequence

The unit occurred over eight class periods and included activities to explore ink diffusing in hot and cold water (Fig. 11.3). Similar to Bielik et al. (2021) and Wilkerson et al. (2015b), students explored the target phenomenon in multiple media using paper modeling and ABM. In our design, however, students validated their models against real-world data tightly integrated within the platform throughout the unit instead of treating external video data as a starting point for their modeling work. Throughout the sequence, students were instructed to focus on how the drop of ink in clear water gradually colored the water and how this related to heat and the interaction of molecules. The classroom teacher implemented the instructional sequence independently with minimal researcher support.

The class met twice a week. To introduce the topic of diffusion, the teacher placed lavender oil on tissue paper and asked students to raise their hands when they smelled the oil. The class briefly discussed the spatial dimensions of the scent's movement through the room. They then followed the sequence described in Fig. 11.3. They conducted an experiment comparing the rate of ink spread in hot and cold water, running three trials for each water temperature and recording their data. They then analyzed the data collaboratively and individually drew paper models to explain the difference in the rate of ink spread across the two conditions. Students were introduced to MoDa through four challenges focused on different programming blocks. Students then worked in pairs to draw another paper model of their explanation before collaboratively coding computational models to explain their observations and compare their models to the video data. Students shared their paper and computational models with the class to get feedback from their peers and the teacher. On the last day, students discussed the validity of their models and watched a video that described the canonical explanation for diffusion.

11.3.4 Data Sources

We analyzed five sources of student data (bolded and italicized in Fig. 11.3):

1. Students' **written responses** to open-ended questions on the pre-survey about how and why ink diffuses in water (Day 0).
2. Students' **individual drawings** of the experiment (Day 2). *Two students were absent on this day.
3. Students' **pair drawings** after first working in MoDa (Day 5).
4. Students' **spoken model share-outs** in pairs as they presented their computer models to their classmates (Day 7).
5. Students' **written responses** to open-ended questions on the post-survey about how and why ink diffuses in water (Day 9).

Collected across the instructional sequence, these five data sources provide a window into how students' mechanistic reasoning and conceptual understanding of diffusion developed throughout the unit. However, because they represent three different modalities—written responses, drawings, and verbal utterances—and diffusion was presented to students as a visuo-spatial phenomenon, we anticipated that these various data sources might surface learners' thinking in different ways. Our data sources were collected over time, with some repetition across the unit, allowing us to pursue this hypothesis.

11.3.5 Data Analysis

Our data include both individual- and pair-level data. Of our five data points, three were at the individual level, while two were at a pair level. To respect variation between students in a pair, we treat the individual as the level of analysis and assign pair-level data to each student in that pair. We acknowledge that in ascribing pair-level data to each individual, this approach would potentially misattribute one student's learning to their partner (overestimation). However, the alternatives—either dropping the two pair-level data points or combining individual data for a pair-wise analysis—give a limited view of the learning trajectories and blur distinctions between students in a pair, respectively.

In coding whether a specific explanation had elements of mechanistic reasoning, we used a modified version of Russ et al.'s (2008) framework (Table 11.1). In doing this, we followed Krist et al.'s (2019) practice of omitting Russ et al.'s (2008) first two code categories since the activity prompts already describe both the target phenomenon ("In which does it spread faster?") and the experimental setup ("A drop of ink is placed in a glass of hot water and a glass of cold water"). Additionally, Russ et al.'s "chaining" code focuses on the time dimension, similar to how a storyboard presents events chronologically. Krist et al.'s "linking" code emphasizes movement across multiple levels of scale, from microscopic entities to macroscopic aggregate phenomena. Since we did not want to focus on time or scale, we created a new code, "causality," in place of and closely related to Krist et al.'s (2019) "linking" and Russ et al.'s (2008) "chaining." "Causality" is the deepest level of mechanistic reasoning in our rubric. It is defined as reasoning about how individual entities, along with their properties and interactions as they propagate across time, scale, or other relevant dimensions, give rise to the aggregate phenomenon. A numerical score for mechanistic reasoning was determined by counting the presence of the following five elements that appeared in students' work, regardless of which specific elements they were. See Table 11.1 for the complete list of codes, their definitions, and examples.

Data were coded at the student-data source level. That is, we coded whether each student demonstrated each element of MR at least once at the time point represented by each data source (not how frequently each student used each MR element at each

Table 11.1 Rubric for mechanistic reasoning about diffusion, adapted from Russ et al. (2008)

Code	Description	Examples of mechanistic reasoning
Identifying entities	Students mention the elements (entities) that play roles in producing diffusion	Water molecules, color molecules, air molecules, particles, elements, and atoms
Identifying activities	Students describe actions and activities that caused diffusion	Molecules spread apart, molecules spread evenly
Identify properties of entities	Students describe the properties (adjectives) of the entities responsible for the target phenomenon	Water molecules are little hard balls that bounce off everything; molecules move faster, and molecules are bigger
Identifying Organization of Entities	Students indicate how the entities are spatially organized and structured	Dye molecules move between the water molecules, cold molecules are closer together, and molecules move from high concentration to low concentration
Causality	Students reason about how the properties and interactions of individual entities give rise to the general phenomenon	The faster the water molecules move, the more they spread the ink

time point). Two researchers independently coded each data source, achieving at least a 91% match rate on a training set by data source and at least an 87% match on the remaining data. The score for work completed by pairs of students (i.e., pair drawings and model presentations) was attributed to both students of the pair.

11.4 Findings

Our findings suggest that students developed mechanistic explanations as they built agent-based models, confirming existing work. We also found that specific elements of mechanistic reasoning (i.e., entities, activities, etc.; see Table 11.1) were expressed with different frequencies across the unit. While the increasing depth of MR is expected over time given its presumed building-block nature, we also found unexpected decreases in expression at certain time points and for certain elements of MR. A deeper analysis suggests that modality (student utterances, drawings, and written responses) likely played a role in shaping which elements of mechanistic reasoning were expressed at different times during the unit.

11.4.1 Students Develop Mechanistic Explanations as they Build Agent-Based Models

Across the instructional sequence, the number of mechanistic reasoning elements that students expressed grew dramatically from a mean of 0.2 elements (out of 5) on the pre-survey to 3.8 elements on Day 7 and 2.8 elements on the post-survey (Fig. 11.4). A one-way repeated measures ANOVA revealed statistically significant differences across time points ($F(4, 15) = 15.3$, $p < 0.001$). Post hoc comparisons using the Tukey HSD test with Bonferroni corrections reveal significant differences between Day 0 and Days 5, 7, and 9, suggesting that students' expressions of MR indeed grew over the course of the unit. Additionally, differences in the means between Day 2 and Day 7 were significant, suggesting that the verbal presentations during Day 7 more clearly indicate growth in students' expressions of MR than at any other time point.

While the mean increases when aggregated across students (Fig. 11.4), individual students show marked variation in the number of MR elements they mention from one day to the next across different media (Fig. 11.5). For example, eight students exhibited a shift of 2 or more units in MR from the pre-survey to the drawing stage (e.g., Joy). Another eight students showed a similar shift from the drawing stage to using MoDa (e.g. student f). Others show an increase in MR elements during the instructional unit but a decrease at the post-test (e.g. Miguel, Tim, and Jade). These various trajectories indicate that students followed individual paths when developing MR while learning about diffusion.

In the next section, we dive into each MR element individually and then explore potential associations with the different representations.

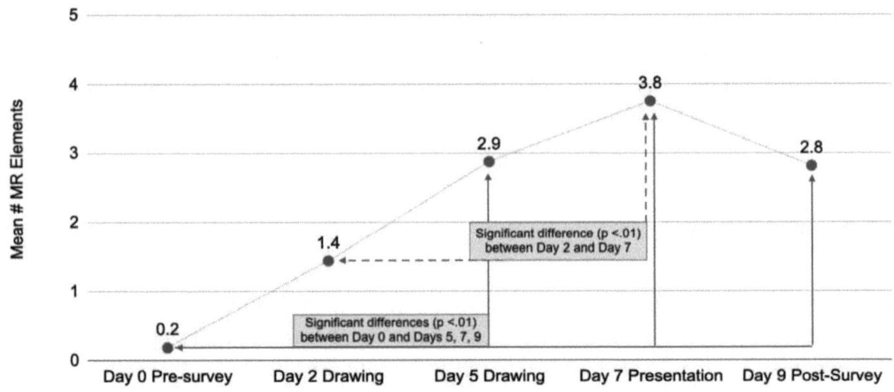

Fig. 11.4 The average number of MR elements mentioned by students

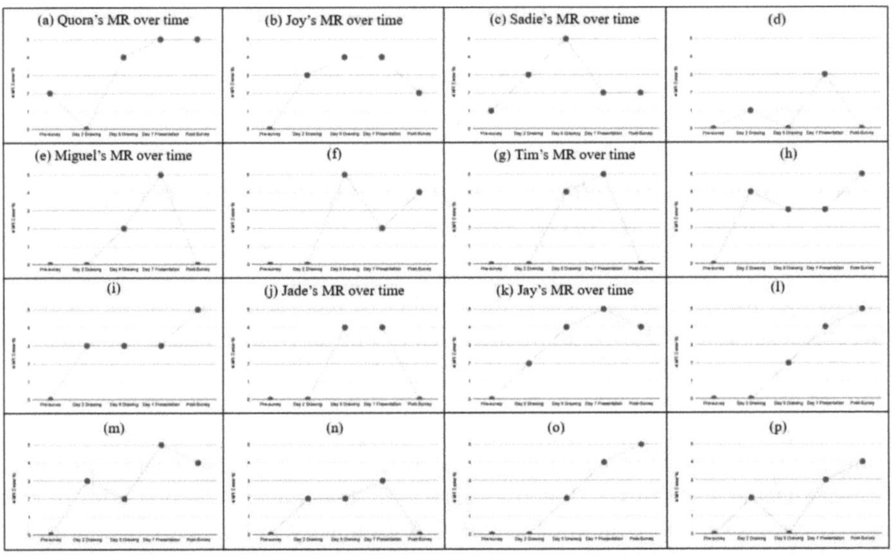

Fig. 11.5 The trajectory of MR elements mentioned by all 16 students over time (pseudo-named focal students are discussed below)

11.4.2 Modality Mediates Students' Expression of Different MR Elements

The findings above considered students' total MR scores as demonstrated at various data collection points throughout the unit. To examine how modality may be shaping how students express MR, we next disaggregate the MR score reported above to examine whether there are differences in the particular elements students demonstrated at each time point (Fig. 11.6).

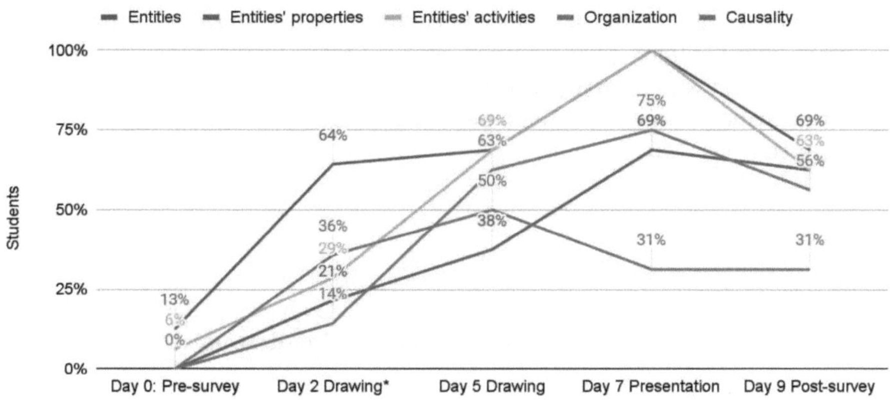

Fig. 11.6 Percentage of students mentioning mechanistic elements in different representations throughout the unit

We find here that most elements of MR follow the more general patterns described above. They all show a general pattern of increase over time as students moved through the unit, suggesting that students did indeed deepen their MR over time as they engaged in ABM activities, and that this change was holistic rather than driven by a subset of elements types. It is worth noting that some elements of MR appeared earlier than others for many students. This included describing entities, which increased by 51% from appearing in 13% of student artifacts in Day 0 to 64% of student artifacts in Day 2, the organization of entities which increased 36%, and describing entities' activities which increased by 23%. These earlier-appearing elements of MR correspond to what are typically described as lower-level elements of MR. Articulations of causality appeared later, toward the middle of the unit. These patterns are consistent with research that highlights the developmental nature of MR–entities and their activities must be identified before their actions and interactions can be considered together to form causal arguments. However, some of the disaggregated patterns are also surprising. For example, more students articulated the *organization* of entities during the drawing activities in Days 2 (36%) and 5 (50%) than at later points in time (31% on Days 7 and 9). We suspect that this is at least partly due to the benefits of drawing as a modality for surfacing the organizational element MR.

We were also surprised to find that, despite an otherwise clear upward trajectory, the proportion of students that demonstrated MR markedly decreased from Day 7 to Day 9, and that this was consistent across nearly all disaggregated elements of MR. Some of the reductions were especially notable: 100% of students articulated system entities and their activities during the Day 7 verbal presentations, but both of these proportions fell below 70% at the Day 9 post-test. As we illustrate in more detail below, we suspect that this is because the post-survey modality of written responses was not especially conducive to articulating key elements of MR, such as entities, properties, and their understanding of cause and effect.

Based on the differential development patterns revealed above and the additional insights yielded by their disaggregation, we next present case studies that examine students' drawings, spoken utterances, and written responses to see how each type of representation may have differentially highlighted (or obscured) different elements of students' MR.

11.4.3 Case Studies

11.4.3.1 Drawings Surface Considerations of Organization

The instructional sequence included two drawings: individual drawings on Day 2 (after running the ink experiment but before using MoDa) and a pair drawing on Day 5 (after a MoDa introduction) to plan their computational models.

Not surprisingly, drawings effectively allowed students to express their ideas about organization, space, and structure of entities involved in diffusion. Identifying the organization of entities as an element of MR occurs when students indicate how the entities are spatially organized and structured. Drawing an explanation of the phenomena on paper requires students to consider spatial and structural features, for example, if and how ink molecules move in between the water molecules, the spacing between molecules, areas where that spacing (concentration) might differ, cold molecules are closer together, and molecules move from high concentration to low concentration. Our data suggest that students' demonstration of organization was higher in the drawings than in other representational media. While 50% of students explicitly featured organizational elements in their drawings, only 31% included this MR element in their verbal presentation or post-test. Next, we give the example of Joy to illustrate this pattern.

Joy, who partnered with Jade, showed no MR elements in her pre-test response ("*it sinks? I am not sure*"). However, she expressed elements of mechanistic reasoning both in her individual and paired drawings (entities, activities, and causality). We focus specifically on her use of organization of entities. When Joy created her first drawing of ink spread in water (Fig. 11.7 panel 1), she wrote, "*I noticed that ink spreads faster in warm water. I think this happened because the adems* [atoms] *spread out in warm water, and the adems* [atoms] *can spread without other adems* [atoms] *in the way. The particles move slower in cold and faster in hot [water].*" In this individual drawing, Joy identified and drew the organization of entities (concentrated vertically in the middle for cold water, spread all over for hot water) and described it in words ("*[atoms] spread out in warm water,*" "*[atoms] can spread without other adems [atoms] in the way*").

Although their second paper model (Fig. 11.7 panel 2) depicted an incorrect theory about how molecules spread ("*cold water molecules freeze, warm does not*"), Joy and Jade demonstrated four out of five elements of mechanistic reasoning, including organization of entities. They both drew and described entities' organization ("*Ink molecules spread open and mixed in with warm water*" "*more bunch*

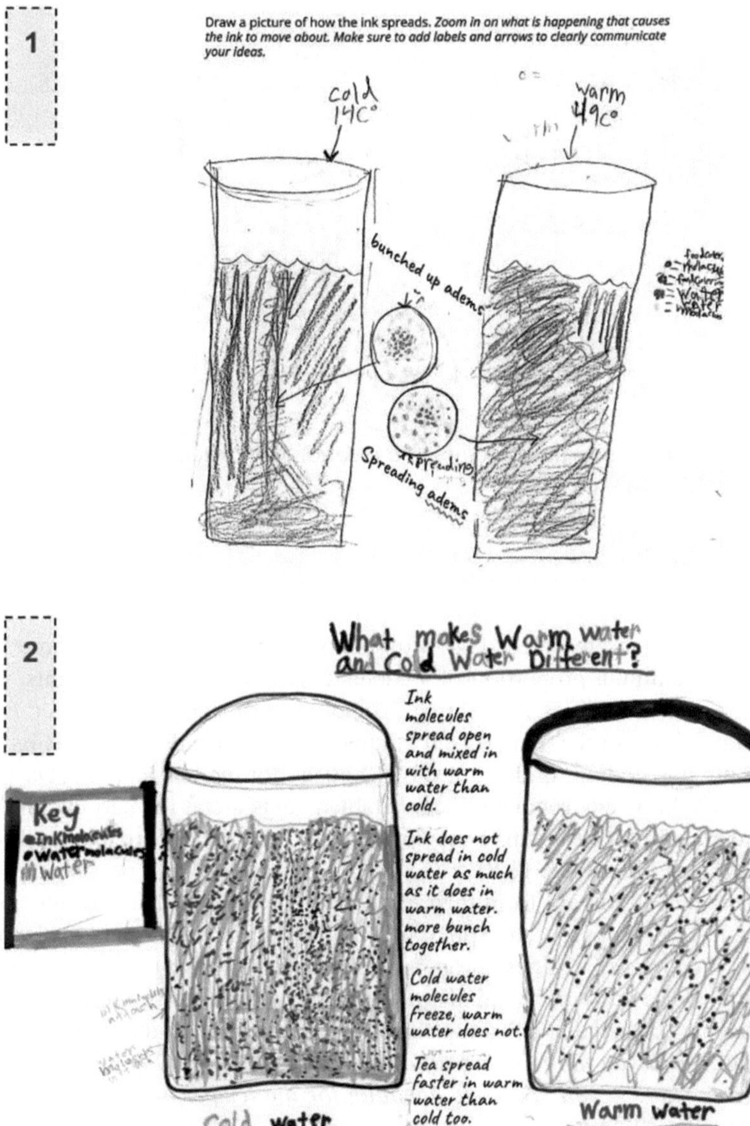

Fig. 11.7 Joy's (1) individual and (2) pair drawing both included organization of entities

together,"), with molecules placed closer together in cold water than in warm. Interestingly, Joy did not mention organization of entities in her verbal presentation or her post-test ("*the ink molecules move faster in water*").

Drawings, pictures, and other symbolic tools are not secondary to language but are crucial elements of students' repertoire for meaning-making and building understanding. In this context, they appeared to help students focus on entities and their

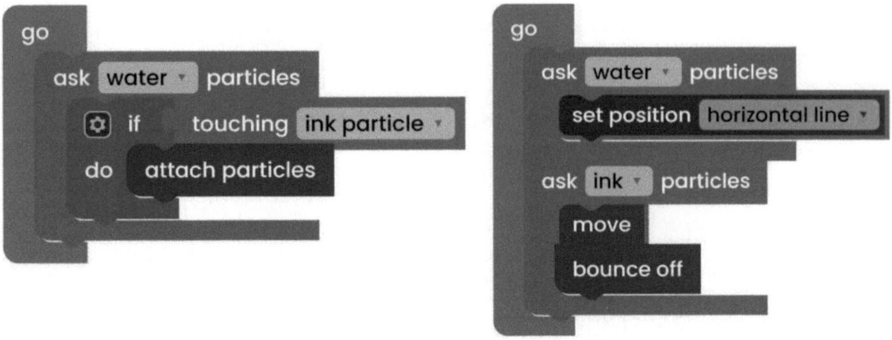

Fig. 11.8 Two examples of students' code with blocks that show the particle level language

organization, examining the structure and location of molecules in relation to other molecules and within the space.

11.4.3.2 Verbal Presentations Surfaced the Most Elements of MR

On Day 7, students presented their MoDa models to the class. This time period marked the maximum proportion of students expressing all MR elements, except for organization of entities (Fig. 11.6). Notably, every student mentioned entities and their activities for the first and only time across our five data points (Fig. 11.6). We attribute this finding to the choice of language for the blocks in MoDa's block library; as an ABM environment, the available agents are at the particle level (i.e., water and ink particles), and the behaviors that students encode are inherently activities of those particles (Fig. 11.8).

Students tended to read from their code when presenting their models, making this particle-level language an explicit part of their presentations and explanations. Students' mentions of entities' properties and causality also peaked during these presentations. We attribute this finding to the teacher's follow-up questions during most students' presentations asking them about disparate results in hot and cold water (e.g., *"Did you notice a big difference between hot and cold water?" "Did that work for hot?" "did you code anything for the temperature?"*) and potential reasons for that difference.

Quora and her partner explored different ideas to explain diffusion. One of their first models, the attaching model, proposed that ink molecules attach to water molecules, causing the ink to spread. The following is their code and share out discussion in the classroom (Fig. 11.9):

When they shared their work with the class on Day 7, Quora and her partner articulated all the MR elements (line 4). They explained that their attaching model didn't match the real-world data from the experiment and video (line 1), specifically because it resulted in "clumps" (line 3) that "*were not what [they] saw in real life and the video*" (line 3) This discrepancy led Quora and her partner to create the

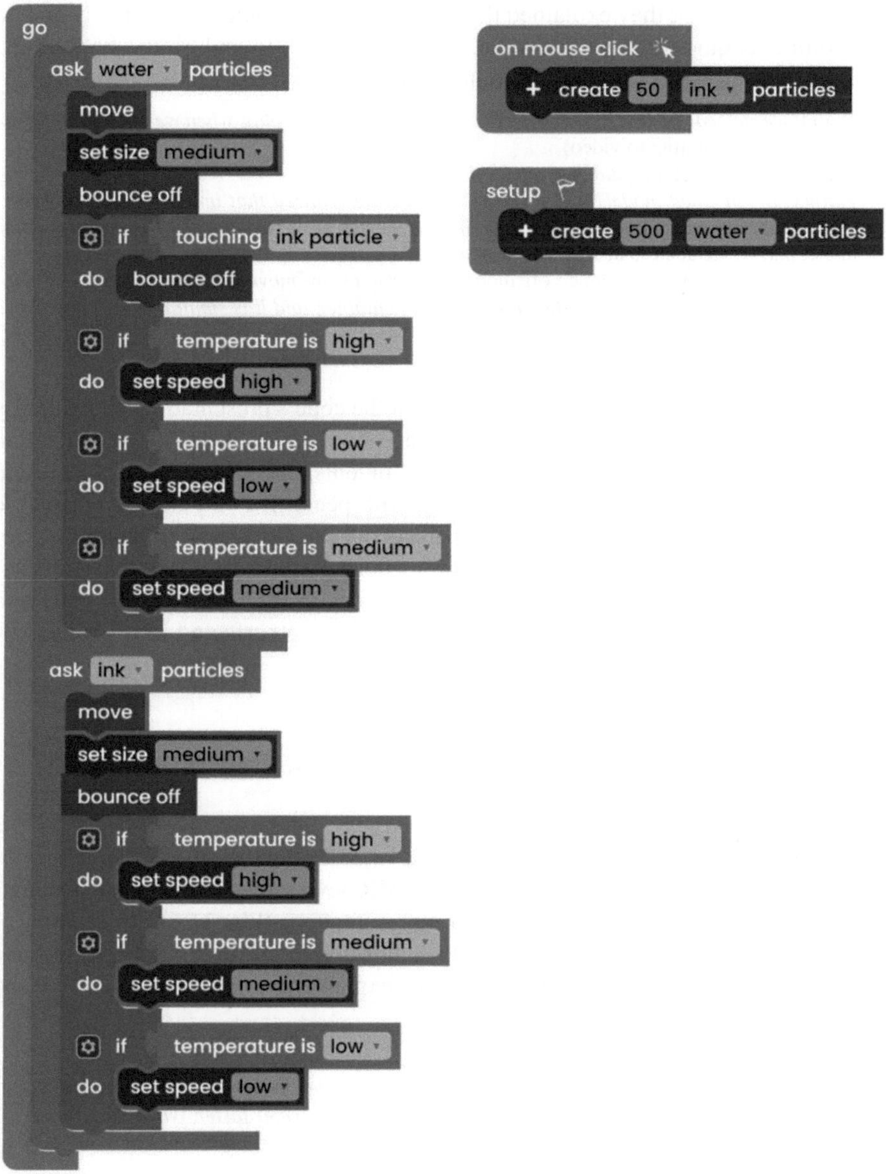

Fig. 11.9 Quora and her partner's MoDa code

"bounce off" model; they explained that in cold water, particles move slowly and do not collide as quickly as in hot water, allowing the ink to sink to the bottom and aligning with the data and video evidence (line 4).

1. Quora: *we started with the attached model like Manny and Nickolas, when we finished it did not look like this* (pointing to video)
2. Teacher: *Can we see the attached model first?*
3. Partner: *it [attached model] did not work because we realized that in the model, there were clumps that were not what we saw in real life and the video. We decided that this was the wrong model and we decided to design the "bounce off" model*
4. Quora: (explaining their bounce off model) *If particles are moving slowly they will hit each other slowly so ink will sink to the bottom of the container, and it is similar in the video when we look at cold water*
   We compared the model with the video, and it looks the same.

In this learning context, the unique qualities of the code's presentation are explicitly verbalized by the students, with some guidance from the teacher. This verbalization is a pedagogical strategy aimed at 6th-grade students, encouraging them to present their work and discuss their understanding with peers. It is a public event whose primary goal is to demonstrate relevant perspectives on different explanations/theories for the phenomenon of diffusion by exploring students' ideas through their MoDa models. The verbal explanation, together with the visual blocks code and scaffolding from the teacher, play an important role in surfacing most of the MR elements (Fig. 11.6).

11.4.3.3 Pre/Post Comparisons Overlook Learning Demonstrated in Other Representation Media

As expected, the pre-test revealed low levels of MR about diffusion. However, we were intrigued that the post-test indicated MR levels lower than what students exhibited earlier in the unit through other representations (Fig. 11.5). An example of the predicted response pattern is Jay, who demonstrated low MR on his pre-test and a high MR on his post-test when asked to explain how ink spreads in hot and cold water:

Pre-test: *I imagine it spreads through the water, maybe because of gravity.*
Post-test: *The ink particles bounce off of the water particles. When the water is warm, the particles move faster, so the ink particles move through the water faster. When the water is cold, the water particles move slower, so when the ink particles bounce off of them, they move slower.*

Within low-MR post-test responses, we identified two categories: students who gave "I don't know" type responses, which may be attributed to test fatigue, and students who gave brief responses that lacked MR elements that these students had previously demonstrated in other representations earlier in the unit.

Figure 11.5 shows seven students whose post-test MR score is 2 or more points below their maximum score (students b, c, d, e, g, j, and n). Table 11.2 gives three of these students' MR scores throughout the unit (Fig. 11.10).

Table 11.2 Examples of students who started and ended with low levels of MR and demonstarted a high-level variation of MR elements within the unit

Student	Day 0 Pre-survey	Day 2 Drawing	Day 5 Drawing	Day 7 Presentation	Day 9 Post survey
Tim (g)	0 (*"It would make its way through the water slowly. I am not sure"*)	0	4	5	0 (*"I don't know"*)
Miguel (e)	0 (*"I don't know what to say"*)	0	2	5	0 (*"by departing in different ways, because that's what I did in a model"*)
Sadie (c)	1 (*"The ink will make its way through the water by interacting with the water molecules? I don't really know…"*)	3	5	2	2 (*"The ink makes its way through the water by diffusion. The speed of the molecules corresponds with the temperature."*)

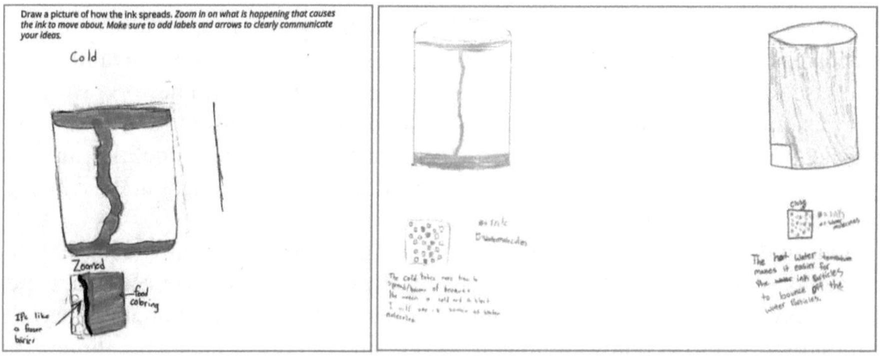

Fig. 11.10 Miguel's individual (left) and pair (right) drawings on Days 2 and 5, respectively

Miguel's learning trajectory started without any MR elements on his pre-test ("*I don't know what to say*"). On Day 2, he drew an artistic depiction of the observed phenomena, which he explained as "*It's like a frozen barrier.*" Even though he drew a zoomed-in picture (underneath his general drawing) that illustrates the water barrier preventing the ink from spreading, he did not use any MR elements. On Day 5, Miguel worked with his partner to create their drawn model of ink spreading in cold and hot water. They added a small illustration for each temperature condition, showing the micro-level of the phenomena, ink and water molecules, and their interaction. They captioned these illustrations: "*The cold takes more time to spread/bounce off because the water is cold, and a block I will use is "bounce off" water molecules. The hot water temperature makes it easier for the ink particles to bounce off the water particles,*" both identifying entities (water molecules, ink particles) and their activities (bounce off). On Day 7, while presenting their code and ideas to the classroom, Miguel and his partner used all five MR elements. While his teacher played their simulation Miguel explained that *"Ink particles are moving down; it spreads down very slowly."* Unfortunately, Miguel's post-test response didn't reflect the MR he demonstrated over the course of the instructional unit. Looking only at his pre- and post-test responses would overlook the MR Miguel and his partner demonstrated in their paired drawing and class presentation.

11.5 Discussion

Existing findings indicate that agent-based modeling (ABM) enhances mechanistic reasoning (MR), and many researchers have shown how students develop their mechanistic reasoning throughout ABM units (Dickes et al., 2016; Eberbach et al., 2021; Louca & Zacharia, 2007; Wilensky, 2003; Wilkerson et al., 2015b; Xiang & Passmore, 2015). These studies often explore students' behavior and how their interaction with learning materials deepens their understanding of scientific phenomena.

In this study, we took the additional step of focusing on how specific elements of mechanistic reasoning differentially appear across different types of modalities they may use to express their learning. Using Russ et al.'s (2008) framework, we carefully examined the components that constitute mechanistic reasoning. We examined mechanistic reasoning (MR) across multiple data sources and modalities, investigating which elements of MR developed, how they evolved throughout the unit, and how they differentially surfaced across activities.

Our findings suggest that certain modalities for sharing MR may reveal certain elements of students' MR more effectively than others. For example, when students draw models on paper individually or with a partner, the element of organization of entities becomes prominent. In our study, more students mentioned the organization of entities in their Day 2 and Day 5 drawings than in their Day 7 verbal presentations or their written Day 9 post-tests (Fig. 11.6). This is consistent with other work that demonstrates visual representations, such as drawing, which may be particularly helpful for students to both develop and share their reasoning about complex mechanisms, spatial relationships, and processes (e.g., Wilkerson et al., 2015a; Ryan & Stieff, 2019). Drawings enable learners to identify the components of a mechanism, such as entities, and are particularly effective at highlighting the organization of these entities on the page. This organizational element is particularly well-suited to drawing tasks and is the most challenging to bring out, with only half of the students mentioning it across all data sources.

Similarly, verbal presentations, often complemented with computer code and supported by a teacher's facilitation, seemed to especially support students in expressing all other MR elements. On average, students' verbal presentations demonstrated the highest rates of all MR elements except for the organization of entities, notably higher than students' written post-survey responses (Fig. 11.6). In fact, half of our participating students exhibited fewer MR elements on their post-test than they did in their verbal presentations (Fig. 11.5). While it could be that certain elements of MR take time to emerge and need more supported engagement with computational modeling, we interpret the overall decline in performance from verbal presentation to post-test as suggesting that writing alone may not capture the full extent of learners' developing MR.

Our findings about the differences in how modality may impact evaluations of students' MR carry the same implication for both researchers and educators. As researchers, we found different levels of MR when we considered students' drawings, verbal presentations, and written responses. We argue that a diversity of modalities can allow for more complete communication of student ideas. In this study, the verbal presentation of computer code seemed particularly effective in surfacing the most complete picture of what students knew about the mechanisms they were studying. From a pedagogical perspective, we suspect that these activities made students' mechanistic thinking and learning visible to them, too. We see this work as aligned with research on the value of external representations as cognitive tools (e.g., Kirsh, 2010).

Our study also supports theory and empirical work that suggests that certain forms of MR might scaffold or precede others (Russ et al., 2008; Krist et al., 2019). It lends further support to pedagogical theories that suggest these developmental trajectories can be supported by well-structured multimodal activities (e.g., Berg & Hultén, 2023; Danish et al., 2020; Wilkerson et al., 2015a). For example, it is reasonable to assume that identifying entities is a prerequisite to describing the organization or causality that underlies a system. Activities or instruction ordered to incrementally target these elements of MR could help more learners develop deeper MR skills.

Like all studies, there are some limitations to this work. Our sample size of 16 is relatively small. However, this size allowed us to dive deeper into the analysis and to identify patterns across all students' learning, an analysis that would have quickly become unwieldy with a larger study population. We also acknowledge that our analysis of both individual and pair-wise data carries inherent trade-offs. This approach may overestimate some students' learning by crediting them for their partners' contributions, but excluding the pair-level data would limit our understanding of learning trajectories across multiple representational media. This study also exemplifies the difficulty in distinguishing, during such a rich pedagogical unit, between pedagogical and evaluative activities and between knowledge expression and development (which are, of course, connected). In terms of analysis, the reliability of assessing students' open-ended drawings, presentations, and written responses can be influenced by several factors. Variability in how students interpret questions, articulate their reasoning, or how evaluators code their responses can impact the reliability of the assessment. To enhance the reliability of the pre- and post-tests, we employed a detailed rubric (see Table 11.1) to standardize the evaluation process. Additionally, we used multiple coders to calculate inter-rater reliability, with the coding conducted by the first two authors of this chapter, both of whom were highly trained in this methodology.

11.6 Conclusions

In conclusion, various non-computational modalities of expression, such as drawing, verbal expression, and writing are commonly used to make claims about student reasoning in ABM studies and curricula. Our study shows that each modality may uncover (or hide) distinct kinds of knowledge essential to mechanistic reasoning. Of all the modalities we examined, students' verbal presentations about their computational code appeared to surface the most profound evidence of mechanistic reasoning. Relying solely on written evaluative representations may provide a weak indication of what students have learned throughout a given unit.

Our data demonstrate that several students received low MR evaluations both in the pre-and post-tests despite otherwise consistently increasing performance throughout the unit. Examining their progress through different modalities, such as drawing or verbal presentations, reveals a higher level of MR. Our more profound analysis of these varying trajectories evidenced both that students developed MR elements at individual rates while learning about diffusion throughout this unit and that the modality of expression also played a crucial role in surfacing this learning. Our findings point to the need for educational designers and researchers to carefully consider the different modalities in the design of units with MR as a key learning goal since the prevalence, combination, and sequencing of these modalities might have significant effects on student learning. As educational researchers, it is vital to recognize that different modalities may reveal specific data, and analysis should incorporate multiple data sources to ensure that results are validated and accurate.

Acknowledgments We thank Ms. Jenny Billings and her students for their thoughtful engagement with this project. This material is based upon work supported by the National Science Foundation under Grant No. DRL-2010413. Any opinions, findings, and conclusions or recommendations expressed in this material are those of the author(s) and do not necessarily reflect the views of the National Science Foundation.

References

Bachtiar, R. W., Meulenbroeks, R. F. G., & van Joolingen, W. R. (2024). Understanding how student-constructed stop-motion animations promote mechanistic reasoning: A theoretical framework and empirical evidence. *Journal of Research in Science Teaching, 61*(2), 289–318. https://doi.org/10.1002/tea.21891

Berg, A., & Hultén, M. (2023). Children's emergent mechanistic reasoning in chemistry: A case study about early primary students' reasoning about the phenomenon of thermal expansion of air. *Chemistry Education Research and Practice, 25*, 92–114. https://doi.org/10.1039/D3RP00169E

Bielik, T., Fonio, E., Feinerman, O., Duncan, R. G., & Levy, S. T. (2021). Working together: Integrating computational modeling approaches to investigate complex phenomena. *Journal of Science Education and Technology, 30*, 40–57. https://doi.org/10.1007/s10956-020-09869-x

Blikstein, P. (2014). Bifocal modeling: Promoting authentic scientific inquiry through exploring and comparing real and ideal systems linked in real-time. *Playful User Interfaces: Interfaces that Invite Social and Physical Interaction*, 317–352. https://doi.org/10.1007/978-981-4560-96-2_15

Blikstein, P., & Wilensky, U. (2009). An atom is known by the company it keeps: A constructionist learning environment for materials science using agent-based modeling. *International Journal of Computers for Mathematical Learning, 14*, 81–119. https://doi.org/10.1007/s10758-009-9148-8

Danish, J. A., Enyedy, N., Saleh, A., & Humburg, M. (2020). Learning in embodied activity framework: A sociocultural framework: A sociocultural framework for embodied cognition. *International Journal of Computer-Supported Collaborative Learning, 15*, 49–87. https://doi.org/10.1007/s11412-020-09317-3

Dickes, A. C., Sengupta, P., Farris, A. V., & Basu, S. (2016). Development of mechanistic reasoning and multilevel explanations of ecology in third grade using agent-based models. *Science Education, 100*(4), 734–776. https://doi.org/10.1002/sce.21217

Eberbach, C., Hmelo-Silver, C. E., Jordan, R., Taylor, J., & Hunter, R. (2021). Multidimensional trajectories for understanding ecosystems. *Science Education, 105*(3), 521–540. https://doi.org/10.1002/sce.21613

Eloy, A., Wolf, J., Fuhrmann, T., Bumbacher, E., Wilkerson, M., & Blikstein, P. (2022). A2S: Designing an integrated platform for computational modeling & data analysis for sustained investigations in science classrooms. In *Proceedings of the 2022 Annual Meeting of the International Society for the Learning Sciences (ISLS 2022)*, Hiroshima, Japan.

Eloy, A., Fuhrmann, T., de Deus Lopes, R., Blikstein, P., & Wagh, A.. (n.d.) *Opening the black box: Student motivations in an unpackable-block-based computational modeling environment* (under review).

Fuhrmann, T., Wagh, A., Eloy, A., Wolf, J., Bumbacher, E., Wilkerson, M., & Blikstein, P. (2022). Infect, attach or bounce off?: Linking real data and computational models to make sense of the mechanisms of diffusion. In Chinn, C., Tan, E., Chan, C., & Kali, Y. (Eds.), *Proceedings of the 16th international conference of the learning sciences – ICLS 2022* (pp. 1445–1448). International Society of the Learning Sciences. https://doi.org/10.22318/icls2022.1445

Fuhrmann, T., Rosenbaum, L., Wagh, A., Eloy, A., Wolf, J., Blikstein, P., & Wilkerson, M. (2024). Right but wrong: How students' mechanistic reasoning and conceptual understandings shift when designing agent-based models using data. *Science Education*, 1–25. https://doi.org/10.1002/sce.21890

Hmelo-Silver, C. E., & Azevedo, R. (2006). Understanding complex systems: Some core challenges. *The Journal of the Learning Sciences, 15*(1), 53–61. https://doi.org/10.1207/s15327809jls1501_7

Hmelo-Silver, C. E., & Pfeffer, M. G. (2004). Comparing expert and novice understanding of a complex system from the perspective of structures, behaviors, and functions. *Cognitive Science, 28*(1), 127–138. https://doi.org/10.1207/s15516709cog2801_7

Horn, M. S., Brady, C., Hjorth, A., Wagh, A., & Wilensky, U. (2014). Frog pond: A codefirst learning environment on evolution and natural selection. In *Proceedings of the 2014 conference on interaction design and children* (pp. 357–360). https://doi.org/10.1145/2593968.2610491

Hsiao, L., Lee, I., & Klopfer, E. (2019). Making sense of models: How teachers use agent-based modeling to advance mechanistic reasoning. *British Journal of Educational Technology, 50*(5), 2203–2216. https://doi.org/10.1111/bjet.12844

Kirsh, D. (2010). Thinking with external representations. *AI & SOCIETY, 25*, 441–454.

Krist, C., Schwarz, C. V., & Reiser, B. J. (2019). Identifying essential epistemic heuristics for guiding mechanistic reasoning in science learning. *Journal of the Learning Sciences, 28*(2), 160–205. https://doi.org/10.1080/10508406.2018.1510404

Langbeheim, E., & Levy, S. T. (2018). Feeling the forces within materials: Bringing intermolecular bonding to the fore using embodied modelling. *International Journal of Science Education, 40*(13), 1567–1586. https://doi.org/10.1080/09500693.2018.1487092

Levy, D. (2013). How dynamic visualization technology can support molecular reasoning. *Journal of Science Education and Technology, 22*(5), 702–717. https://doi.org/10.1007/s10956-012-9424-6

Louca, L. T., & Zacharia, Z. C. (2007). The use of computer-based programming environments as computer modelling tools in early science education: The cases of textual and graphical program languages. *International Journal of Science Education, 30*(3), 287–323. https://doi.org/10.1080/09500690601188620

Louca, L. T., Zacharia, Z. C., & Constantinou, C. P. (2011). In Quest of productive modeling-based learning discourse in elementary school science. *Journal of Research in Science Teaching, 48*(8), 919–951. https://doi.org/10.1002/tea.20435

Machamer, P., Darden, L., & Craver, C. F. (2000). Thinking about mechanisms. *Philosophy of Science, 67*(1), 1–25. https://doi.org/10.1086/392759

MIT Scheller Teacher Education Program. StarLogo Nova [Computer Software]. (2019). Available online: https://education.mit.edu/project/starlogo-nova/. Accessed August 2024.

Perkins, D. N., & Grotzer, T. A. (2000). Models and moves: Focusing on dimensions of causal complexity to achieve deeper scientific understanding. In *Paper presented at the annual conference of the American Educational Research Association*, New Orleans, LA.

Pluta, W. J., Chinn, C. A., & Duncan, R. G. (2011). Learners' epistemic criteria for good scientific models. *Journal of Research in Science Teaching, 48*(5), 486–511. https://doi.org/10.1002/tea.20415

Repenning, A. (2017). Moving beyond syntax: Lessons from 20 years of blocks programing in AgentSheets. *Journal of Visual Languages and Sentient Systems, 3*(1), 68–91. https://doi.org/10.18293/VLSS2017-010

Russ, R. S., Scherr, R. E., Hammer, D., & Mikeska, J. (2008). Recognizing mechanistic reasoning in student scientific inquiry: A framework for discourse analysis developed from the philosophy of science. *Science Education, 92*(3), 499–525. https://doi.org/10.1002/sce.20264

Ryan, S. A. C., & Stieff, M. (2019). Drawing for assessing learning outcomes in chemistry. *Journal of Chemical Education, 96*(9), 1813–1820. https://doi.org/10.1021/acs.jchemed.9b00361

Saba, J., Hel-Or, H., & Levy, S. T. (2021). Much. Matter. In *Motion: Learning by modeling systems in chemistry and physics with a universal programming platform.* Interactive Learning Environments, pp. 1–20. https://doi.org/10.1080/10494820.2021.1919905

Salmon, W. C. (1979). Why ask, 'Why?'? An inquiry concerning scientific explanation. In *Hans Reichenbach: logical empiricist* (pp. 403–425). Springer Netherlands. https://doi.org/10.1007/978-94-009-9404-1_15

Sengupta, P., Dickes, A., & Farris, A. V. (2021). *Voicing code in STEM: A dialogical imagination.* MIT Press.

Shiroda, M., Franovic, C. G., de Lima, J., Noyes, K., Babi, D., Beltran-Flores, E., Kesh, J., McKay, R. L., Persson-Gordon, E., Cooper, M. M., Long, T. M., Schwarz, C. V., & Stoltzfus, J. R. (2024). Examining and supporting mechanistic explanations across chemistry and biology courses. *CBE—Life Sciences Education, 23*(3), ar38. https://doi.org/10.1187/cbe.23-08-0157

Springer, K., & Keil, F. C. (1991). Early differentiation of causal mechanisms appropriate to biological and nonbiological kinds. *Child Development, 62*(4), 767–781. https://doi.org/10.1111/j.1467-8624.1991.tb01568.x

Wagh, A., & Wilensky, U. (2018). EvoBuild: A quickstart toolkit for programming agent-based models of evolutionary processes. *Journal of Science Education and Technology, 27*(2), 131–146. https://doi.org/10.1007/s10956-017-9713-1

Wagh, A., Fuhrmann, T., Eloy, A. A. D. S., Wolf, J., Bumbacher, E., Blikstein, P., & Wilkerson, M. (2022, June). MoDa: Designing a tool to interweave computational modeling with real-world data analysis for science learning in middle school. In *Proceedings of the 21st annual ACM interaction design and children conference* (pp. 206–211). https://doi.org/10.1145/3501712.3529723.

Wilensky, U. (1999). *NetLogo (and NetLogo user manual).* Center for Connected Learning and Computer-Based Modeling, Northwestern University. http://ccl.north-western.edu/netlogo

Wilensky, U. (2003). Statistical mechanics for secondary school: The GasLab multi-agent modeling toolkit. *International Journal of Computers for Mathematical Learning, 8*, 1–41. https://doi.org/10.1023/A:1025651502936

Wilensky, U., & Resnick, M. (1999). Thinking in levels: A dynamic systems approach to making sense of the world. *Journal of Science Education and Technology, 8*, 3–19. https://doi.org/10.1023/A:1009421303064

Wilkerson, M. H., Gravel, B. E., & Macrander, C. A. (2015a). Exploring shifts in middle school learners' modeling activity while generating drawings, animations, and computational simulations of molecular diffusion. *Journal of Science Education and Technology, 24*, 396–415. https://doi.org/10.1007/s10956-014-9497-5

Wilkerson, M., Wagh, A., & Wilensky, U. (2015b). Balancing curricular and pedagogical needs in computational construction kits: Lessons from the DeltaTick project. *Science Education, 99*(3), 465–499. https://doi.org/10.1002/sce.21157

Wilkerson, M., Shareff, B., Gravel, B., Shaban, Y., & Laina, V. (2017). Exploring computational modeling environments as tools to structure classroom-level knowledge building. In B. K. Smith, M. Borge, E. Mercier, & K. Y. Lim (Eds.), *Making a difference: Prioritizing equity and access in CSCL, 12th international conference on computer supported collaborative learning (CSCL) 2017* (Vol. 1). International Society of the Learning Sciences.

Xiang, L., & Passmore, C. (2015). A framework for model-based inquiry through agent-based programming. *Journal of Science Education and Technology, 24*, 311–329. https://doi.org/10.1007/s10956-014-9534-4

Part IV
Contribution of Mechanistic Reasoning to Scientific Literacy

Chapter 12
Mechanistic Reasoning In-School Versus Mechanistic Reasoning In-Life

Ryan L. Stowe, Rosemary S. Russ, and Lindsay B. Wells

12.1 Introduction

Recent reforms in science education have centered our field's attention on the need to move beyond accumulation of science knowledge and toward engaging in scientific practices (Schwarz et al., 2016; The National Research Council, 2012). The idea—now nearly universally accepted in standards (NGSS Lead States, 2013), curricula (Cooper & Klymkowsky, 2013; Shwartz et al., 2008), and assessments (National Research Council, 2013)—is that students need to know not only what scientists know but also what scientists do to construct that knowledge. Underlying this idea is the assumption that if students can construct and critique knowledge in the way that scientists do, then they can be successful at navigating issues of a scientific nature that emerge in their daily lives (The National Research Council, 2012).

One scientific practice that has received substantial attention in the literature is that of developing and using models (Passmore et al., 2014; Schwarz et al., 2009). The Next Generation Science Standards (NGSS) describes this practice as a way to make sense of scientific phenomena, processes, or relationships (NGSS Lead States, 2013). For example, to explain why they can smell paint at a distance, students might construct and use representations of particles traveling from the paint bottle to their nose. Inherent in NGSS's view of models is the notion that these models are causal—they describe scientific phenomena in terms of how one object or process leads to a change in some other object or process.

To highlight the causal nature of modeling, here we explore the related scientific activity of mechanistic reasoning. Russ et al. (2008) proposed that mechanistic reasoning is an important part of school science in that it allows students to draw on

R. L. Stowe (✉) · R. S. Russ · L. B. Wells
Department of Chemistry, University of Wisconsin – Madison, Madison, WI, USA
e-mail: rstowe@chem.wisc.edu

N. Graulich et al. (eds.), *Exploring Mechanistic Reasoning in Science Education*, Contributions from Science Education Research 17,
https://doi.org/10.1007/978-3-032-10246-1_12

their existing resources for causal reasoning to build robust accounts of phenomena consistent with those constructed by professional scientists. In that work, the authors defined mechanistic reasoning as made up of a variety of steps from Describing Target Phenomena, to Identifying Entities and their Properties, to Chaining together multiple steps in a causal process.

In the time that has elapsed since Russ et al. (2008)'s definition of mechanistic reasoning, many scholars have explored student capabilities for such reasoning (Dood et al., 2020; Russ et al., 2009) and designed curricula and classroom practices to support it (e.g., Bachtiar et al., 2021; Cooper et al., 2016). This and related work on mechanistic reasoning tacitly assumes that such reasoning is inherently valuable. Underlying this assumption are two justifications. First, mechanistic reasoning is inherently good because it resembles the way scientists reason about the physical and natural world (Glennan, 2002; Machamer et al., 2000; Westfall, 1986). Second, as Russ (2014) argues, it is good irrespective of what scientists do because it is a productive way to construct knowledge about the world around us (Ahn & Kalish, 2000; Koslowski, 1996). The latter argument echoes claims elsewhere that science education prepares students to reason about science in their everyday lives (Rudolph, 2023; The National Research Council, 2012).

In this chapter, we unpack the latter assumption about the value of mechanistic reasoning in school science. Specifically, we question the claim that mechanistic reasoning in-school prepares learners for mechanistic reasoning in-life. Instead, we suggest that mechanistic reasoning as it has been defined in the literature, and realized in classrooms, does not resemble how individuals and communities productively construct knowledge in their everyday lives.

12.2 Flint Water Crisis

To contextualize our arguments, we focus attention on the Flint water crisis that began in 2014 and continues to affect residents of Flint, Michigan. Specifically, we will contrast school activities contextualized by the crisis in Flint with ways in which members of the Flint community constructed and critiqued mechanisms to hold state regulators accountable. To enable this discussion, we must first sketch the contours of the Flint water crisis.

12.2.1 The Water Switchover

Prior to 2014, Flint received its water from the Detroit Water and Sewer Department (DWSD). The DWSD drew water from a massive nearby lake (Lake Huron) and treated this water at a plant near the shoreline. Treated water was then pumped through a series of pipelines to the people of Flint. Water from the DWSD was both safe and expensive. Flint was the farthest flung of the communities the DWSD served, and the DWSD charged more for delivering water further distances. The city, saddled with managing ancient and sprawling infrastructure with decreasing tax revenue, was motivated to find a cheaper source of water (Clark, 2018).

The Karegnondi Water Authority (KWA) was billed as an answer to Flint's water cost woes. The KWA would pump raw water from Lake Huron to the communities it served, and this water would be treated at local plants. Moving to KWA-supplied water was sold as a cost-savings measure and approved by both the emergency manager and city council. However, the KWA was merely an idea in the mid-2000s; no pipes were yet constructed and it would be several years before the new infrastructure was ready to distribute lake water to Flint and surrounding communities (Clark, 2018). Additionally, Flint's local water treatment plant had been idle since the 60 s and needed re-tooling to treat raw lake water. Renovations to this treatment plant, according to Flint's utilities administrator at the time, would take some time (Fleming, 2018). Flint was not ready to treat and distribute raw water.

Despite the misgivings of Flint's utilities administrator, the decision was made to temporarily draw raw water from the nearby Flint River, and treat it locally, rather than continue to make use of the DWSD's service while KWA infrastructure was built. In the spring of 2014, Flint shut off the water supply from the DWSD and began using river water (Clark, 2018).

12.2.2 *Toxic Water*

Flint residents soon began noticing something off about their water. Bethany Hazard, a retired cancer survivor, noticed that water from her tap seemed murky and foamy. She elected to buy bottled water rather than consume what flowed from her faucet. Lathan Jefferson contacted the Environmental Protection Agency (EPA) office in Chicago to report rashes he believed were due to changes in his water. These experiences were common enough that local news reported, approximately a month after the water switchover, that residents were "avoiding the tap" and "drinking bottled water instead" (Clark, 2018, p 33).

As it happens, there was a serious problem with the water in Flint in the summer of 2014. The new water treatment program implemented by the city did not include corrosion control. The Michigan Department of Environmental Quality (MDEQ) told staff at the treatment plant that such measures were not necessary at the present time (Flint Water Advisory Task Force, 2016). Corrosion control keeps the metal pipes in America's vast water infrastructure from degrading. Specifically, chemical species such as orthophosphates react with metals in pipes to form a mineral-like crust that prevents the pipes from leaching metal into the water. In the absence of corrosion control, metals like lead are able to dissolve in water flowing through the pipes and be distributed to residents all over the city (Clark, 2018).

Lead is a well-known toxin that is especially harmful to children. It accumulates in teeth, bone, and soft tissues and can cause a broad spectrum of issues (e.g., brain swelling, anemia, abdominal pain, kidney failure, reduced attention spans, seizures, coma; Clark, 2018). There is no safe level for lead in water. As such, when it was found that LeeAnn Walter's water contained alarmingly high levels of lead, and that members of her family had correspondingly high concentrations of lead in their bloodstream, concern was more than warranted. Over 270 samples from around the city, carefully collected by residents of Flint in collaboration with researchers from Virginia Tech, showed that LeeAnn's experience was not an aberration. Samples had an average of 26.79 parts per billion of lead in them, nearly three times the safety standard set by the World Health Organization (10 ppb; Clark, 2018). The highest tested sample came in at 1051 ppb (Edwards et al., 2015).

This is the crux of the Flint water crisis, which poisoned a city and took heroic efforts from the Flint community, tenacious reporters, and collaborating scientists to resolve (at least insomuch as the crisis is resolved). We will discuss the ways in which mechanistic reasoning was used by members of the Flint community to legitimize their experiences in Sect. 4. First, however, we turn to school science, which has produced a number of chemistry activities contextualized by the Flint water crisis (e.g., Buckley & Fahrenkrug, 2020; Terry, 2017; Yu & Linden, 2022). In unpacking one of these activities, we ask: what aspects of mechanistic reasoning does this activity communicate as most valuable? We will then explore how/whether mechanistic reasoning emphasized in school maps onto mechanistic reasoning engaged in by members of the Flint community.

12.3 Mechanistic Reasoning in-School

The Flint water crisis—with its contested knowledge about chemicals both in the water from the pipes and in the bodies of members of the surrounding community—seems to be precisely the sort of science that people encounter in their daily lives for which science education could be useful. It shares several features with topics like climate change or vaccination that are often cited as socio-scientific issues that science education can prepare people to solve. Specifically, like climate change or vaccination, the Flint water crisis involves a relatively limited amount of scientific knowledge and is seen as inherently interesting because it has a substantial impact on how people live and thrive (or not) in the world. Given these features, it is not surprising that the American Chemical Society's ChemMatters magazine states that:

> Through knowledge of chemistry, including the formation of precipitates, the pH scale, and the redox chemistry of half-reactions, scientists were able to solve the Flint water mystery. The crisis was stopped before it became worse, and Flint's water supply was made safe—all thanks to chemistry. (Dingle, 2016, p 8)

This quote suggests that it is chemistry knowledge—precisely of the sort that can be taught and explored in school science classrooms—that "saved the day" for the residents of Flint. As a result, it implies that if lay people could just understand and engage in science *as it is taught in school* then they could solve problems such as this in their lives.

Given claims like the one above, it is not surprising that several different school activities have been designed around "solving" the Flint water crisis (Buckley & Fahrenkrug, 2020; Terry, 2017; Yu & Linden, 2022). To see what sorts of science learning students engage in during these activities, we focus our attention on a high school chemistry lesson meant to introduce the topic of chemical reactions (Terry, 2017). We chose this lesson because it encourages students to engage in mechanistic reasoning to support their understanding of the Flint water crisis. The features of school mechanistic reasoning we pull from this lesson are shared by other high school and college chemistry lessons contextualized by the crisis in Flint (e.g., Buckley & Fahrenkrug, 2020; Yu & Linden, 2022).

Students are meant to begin learning about chemical reactions in the context of the Flint water crisis by reading their textbook and a news article about the chemical processes involved in the contamination of Flint's water due to corrosion in the metal pipes (Torrice, 2016). In the lesson, students are led through exercises and content about various types of chemical reactions and solubility rules. They are then given a variety of compounds that were present in the Flint water pipes and water both before and after 2014 and asked to predict the precipitates for the different chemical reactions that occurred.

With all of this chemical knowledge, students are then asked to reason mechanistically by balancing chemical equations and assessing the relative amount of different compounds present in the water. In the language of Russ et al. (2008)'s coding scheme for mechanistic reasoning, students identify several *entities* involved in the process including PO_4^{3-}, Pb, OH^-, Fe, and Cl_2. They are further asked to reason about the *properties of those entities*—namely their net charges, and associated ions. These entities and their properties are *organized* in differing relative amounts inside the pipe and water that flowed through it. Finally, students *chain* those entities with those properties in that organization together using solubility, solution pH, and electron exchange as the *activities* to produce the following mechanism, where "→" means "leads to":

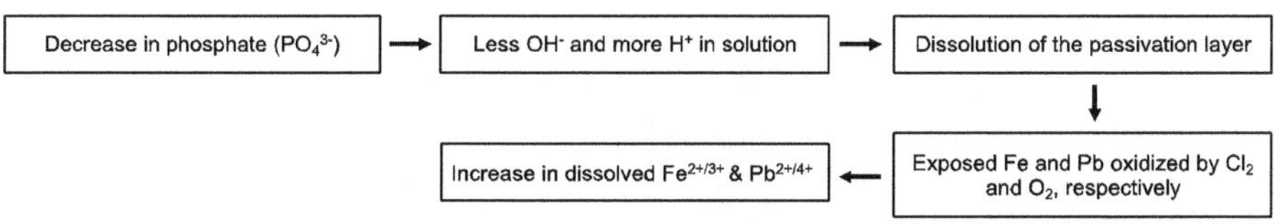

Images of the stages in this mechanism (what some might call models) are presented both in the lesson and in the news article (Terry, 2017; Torrice, 2016). Those images (see Fig. 12.1) show students the entities and organization as well as the processes by which lead entered water from the pipes in Flint after 2014.

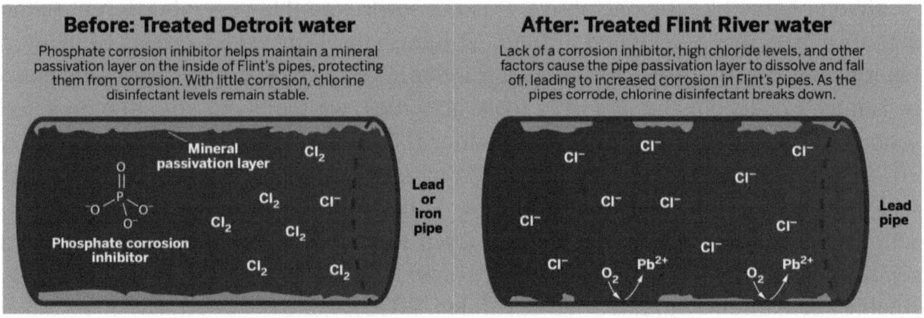

Fig. 12.1 Snapshots of a mechanism explaining high levels of lead in Flint water. (Reprinted with permission from *Chemical & Engineering News,* copyright © 2016 the American Chemical Society. This article was first published on Feb. 11, 2016 in volume 94, issue 7)

By understanding this mechanism and making predictions about its outcome, students are directly shown how the water in Flint became toxic and harmful to its residents. In doing so, this lesson represents a robust way to integrate chemistry and mechanistic reasoning into the exploration of the injustice of the Flint water crisis.

However, what we also see in this lesson is that the mechanistic reasoning about chemistry that students do is only valued insofar as it is correct. The questions at the end of the lesson (e.g., What led to the dissolution of the passivation layer, exposing of the metal (Pb and Fe) pipes?) ask individual students to select mechanistic answers (e.g., If X, then Y) but they are only given credit for those answers if they match the canonical mechanism presented in class. This suggests an "all or nothing" approach to mechanistic reasoning in which any deviation from normative mechanistic connections indicates that students fail to understand the chemical reactions that led to the Flint water crisis. As its designers describe:

> The goal [of this lesson] is to have students learn how to identify [oxidation/reduction, acid/base, and precipitation] reactions and understand their importance in a real-world setting with significant consequences. (Terry, 2017)

This quote demonstrates that the goal here is to center the content of chemistry—the various rules and reactions that gave rise to the problems in Flint. While the designers certainly encourage the use of mechanistic reasoning to understand that content—the content itself is what is most important in the lesson.

We propose that, although this lesson asks students to engage in sophisticated mechanistic reasoning (as defined by Russ et al. (2008)) about a socioscientific issue, its ultimate goal of having students reach canonical correctness through canonically correct (and pre-specified) mechanistic components limits its potential for supporting students in using mechanistic reasoning in their daily lives. Specifically, this lesson allows individual students to reason about only a very narrow set of *entities*, *properties*, *organizations*, and *activities* that are provided by the teacher. Only those mechanistic components introduced by the teacher or the reading are available for students to consider in their mechanistic reasoning. Students have no agency in investigating or determining possible mechanistic components to include. Further, that reasoning occurs in the service of understanding something that has already been determined by an authority.

Given that this lesson supports a relatively constrained form of mechanistic reasoning in which students attempt to match their knowledge to existing, established knowledge, we can ask ourselves whether and how that type of reasoning might be useful for explaining something in daily life for which the answer is not known. We suggest that this pattern-matching-style mechanistic reasoning does little to prepare students to grapple with complex and often unbounded problems such as those faced by the community in the Flint water crisis. In what follows we articulate how the Flint community grappled with mechanisms and demonstrate that this grappling bears little resemblance to the kind of pattern-matching students engage in during this lesson.

12.4 Mechanistic Reasoning In-Flint

The concerns about water safety voiced by LeeAnn Walters, Lathan Jefferson, and many other Flint residents were not initially taken seriously by those with authority. Contrary to what the ACS and the examined school activity suggests, stringing together science facts into a normative mechanism often does not immediately win the day. Here, we will describe instances where some collection of Flint residents, scientists, and authorities critiqued, negotiated, and constructed working mechanisms related to the water in Flint. In doing so, we aim to illustrate that how and why this work was undertaken is distinctly different from what is typically done in schools. Figure 12.2 shows how the events we discuss in this chapter are temporally related. This timeline is far from comprehensive. Those who would like a more complete picture of the Flint water crisis are advised to read Anna Clark's *The Poisoned City* (Clark, 2018).

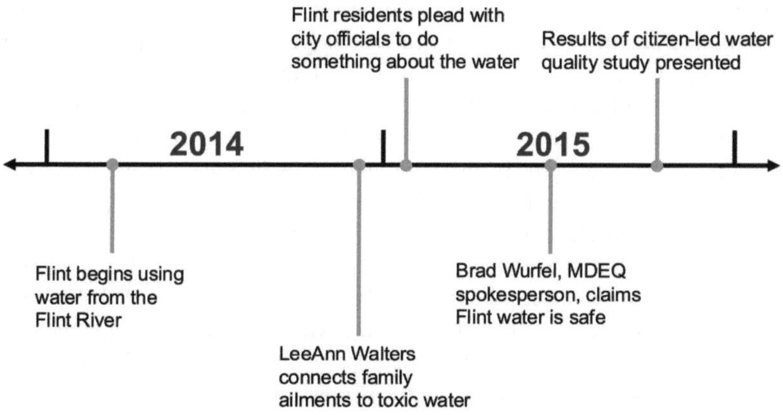

Fig. 12.2 Timeline illustrating the temporal relationship between episodes of the Flint water crisis discussed in this chapter

12.4.1 Critiquing Officials' Mechanisms

Flint residents' mechanistic reasoning started with considering the reasonableness of official mechanistic accounts. State and local officials in Flint claimed that water taken from the Flint river and treated locally was just as safe as Lake Huron water treated and transported by the DWSD (Clark, 2018). This suggests that the composition of water from the two sources is similar (and similarly safe) at the tap. Mechanistic reasoning underlying official claims would be something like: water flows from a natural source to a treatment plant, harmful things are removed and good things are added. This water is then distributed via pipes to various sectors of the city and does not appreciably change during this journey—the pipes do not add or subtract anything of consequence from the water as it flows through. One can then safely consume and use this water.

Two phenomena are the focus of mechanistic reasoning of this sort: water treatment (i.e., removing bad things and adding good things), and water transit from the treatment plant to the tap. In both cases, we can think about lower-scale entities involved (or not) in the phenomenon—these include water particles, particles of good or bad things (which become better specified as the crisis unfolds), and particles that make up the various pipes that comprise Flint's water infrastructure. These entities might take part in a variety of activities—they might be removed (bad particles), added (good particles) or be unaffected by the passage of water (pipe particles). Coordination of entity activities in a particular direction lets us "run" the mechanism (i.e., chaining).

By early 2015, many Flint residents were not buying the tacit mechanism underlying official claims of water safety—these claims were inconsistent with residents' lived experiences. For example, LeeAnn Walters began observing rashes on herself and her family shortly after the water switchover. These became worse over time. By fall of 2014, a whole host of ailments beset the Walters family, including hair loss, cognitive impairment, and abdominal pain. LeeAnn began connecting these symptoms to the water when, near Christmas, brown water started spurting from her tap (Clark, 2018). At around the same time, city officials notified residents that water contained unsafe concentrations of carcinogenic trihalomethanes. This notice stated that steps were being taken to fix this problem at the local treatment plant and "this is not an emergency" (City of Flint, 2015). Despite these official assurances, it was becoming clear to many Flint residents that *something* had changed about the water coming from their tap. In the language of mechanistic reasoning, we could say that the residents chained backwards from their health problems, to the water, to the pipes and infrastructure that held and moved the water.

The people of Flint told those in power about the issues with their water. For example, at a rowdy meeting in January 2015, residents demanded that the city's emergency manager account for the brown, toxic water flowing into their homes. These concerns were summarily dismissed. LeeAnn got city officials to inspect her water only after she presented a doctor's note about how badly her immune-compromised son was reacting to the water. The results of this inspection showed extremely high levels of lead and iron in the water. Rather than use this as evidence to revise their tacit mechanism of water treatment and transport throughout the city, officials constructed an alternative mechanism and attributed LeeAnn's issues to faulty indoor plumbing and agreed to replace the service line that connected her house to the water main (Clark, 2018). The implication here was that LeeAnn's issues were a one-off and the water infrastructure overall was basically fine.

12.4.2 Epistemic Injustice

The January 2015 meeting in city hall is an example of when city officials undermined residents' credibility as knowers and shapers of knowledge. Stated differently, we might say that interactions at this meeting were instances of epistemic injustice (Fricker, 2007). Rather than use residents' lived experiences as an outcome that required drastic changes to their existing mechanism and an impetus for widespread investigation of Flint's water quality, officials opted to dismiss concerns as uncredible—LeeAnn recalls the emergency manager telling her the brown water in the bottle she brought was, "not your water." If some subset of knowers in a community are regularly silenced, that community may "develop epistemological resistance in that these communities do not allow the knowledge and ideas of those perceived as 'non-knowers' to shift the collective perspectives and understandings" (Miller et al., 2018, p 1068). Those with power clearly developed epistemological resistance to the residents of Flint; although the residents appropriately questioned the mechanisms being presented to them, officials rejected their concerns.

Epistemic injustice is a consistent thread throughout the Flint water crisis. Time and again, those with authority undermined community members' status as knowers. In response to the epistemic injustices they experienced, residents collaborated with scientists and journalists to refine and elaborate on their mechanistic accounts—with the goal of having their lived experiences recognized and legitimized by those in power. Therefore, a mechanism was useful if it led to further investigations and, ultimately, to changes in water quality. Stated more generally, mechanistic reasoning during the Flint water crisis was undertaken with the goal of civic action, not for the purpose of matching some predetermined, expert-like drawing on an answer key.

The different goals motivating mechanistic reasoning in-class and in-Flint are likely to manifest in different ways of thinking and acting. Students engaged in the high school activity mentioned earlier would revise their reasoning only if their knowledge product conflicted with the authority-defined answer. By contrast, Flint residents continued to revise their own mechanism (e.g., by further specifying properties of entities, by checking consistency with available evidence) until it served to persuade authorities there was a crisis that demanded immediate action. We illustrate this sort of mechanism refinement below with a concrete example of residents collaborating with Virginia Tech scientists.

12.4.3 *Elaborating Mechanisms for the Purpose of Legitimizing Lived Experiences*

Let me start here—anyone who is concerned about the lead in the drinking water in Flint can relax.

Brad Wurfel, Michigan Department of Environmental Quality spokesperson, July 2015 (Clark, 2018, p 118)

In mid-2015, city and state officials buttressed their claims that "everything is fine" with the results of a 169-sample test of lead levels across the city. These tests showed an average lead level of 11 ppb, which is within acceptable limits. There were many issues with this study, including omission of two data points showing high lead levels, the practice of pre-flushing water lines prior to testing, and concentrating sampling around areas with newer water infrastructure (Clark, 2018). Additionally, claims made by officials were not consistent with the lived experiences of Flint residents like LeeAnn Walters.

To investigate the reasonableness of the official "everything is fine narrative", Flint residents built connections with scientists who had experience investigating drinking water safety. Supported by a grant from the National Science Foundation, Marc Edwards of Virginia Tech and his graduate students worked with the Coalition for Clean Water—a community advocacy group formed in 2015—to collect and analyze hundreds of water samples from around Flint. The testing protocol used in this study remedied many of the issues with the earlier city tests. For example, lines were not pre-flushed, water was to be collected from the cold tap at high flow, and multiple samples were collected from each site. Analysis of 277 community-collected samples revealed an average of 26.79 parts per billion of lead—almost three times the WHO safety standard (Clark, 2018; Edwards et al., 2015). Contaminated water was found throughout the city, validating the residents' understanding that LeeAnn's experience was not a one-off.

The results of the Virginia Tech study allowed Flint residents to elaborate their mechanistic accounts of how their water became toxic. "Good things" (i.e., corrosion control) were not added as part of water treatment and so "bad things" (i.e., lead, iron) were able to leach into the water. There are a few key characteristics of the resident's mechanistic reasoning that are worth highlighting here. First, this reasoning was not an individual activity but rather one that involved a community of concerned residents making connections with knowers whose expertise was thought useful to that community (Virginia Tech scientists). This is what Feinstein and Baram-Tsabari would call "curating an epistemic network" where an "epistemic network" is "a set of people who support sensemaking by providing new information and by aiding in the interpretation and reconstruction of scientific knowledge in context" (Feinstein & Baram-Tsabari, 2024, p 2050). Second, the purpose of elaborating the community's mechanistic account was to prompt action from those with authority. Specifically, members of the Coalition for Clean Water sought a return to the DWSD water supply. A mechanism achieved its purpose if it moved officials toward taking meaningful action.

12.5 How Can We Support (Potentially) Useful Mechanistic Reasoning?

We claim that, to support learners in using mechanistic reasoning in their post-school daily lives, mechanistic reasoning emphasized in-class should more closely resemble mechanistic reasoning useful in-life. That is, student communities should construct and critique mechanisms for reasons that are sensible beyond the bounds of the classroom. We suspect few would disagree with this intuitively plausible premise. Unfortunately, many standards documents (NGSS Lead States, 2013), curricula, and research projects (Dood et al., 2020) emphasize a vision of mechanistic reasoning in which individual students are to string together science facts and pictures to construct normative mechanisms for the purpose of pleasing the teacher. This sort of activity is meaningfully different from the ways in which communities construct and use mechanisms to promote just outcomes. Specifically, as we see in the cases of Flint residents' mechanistic reasoning, communities tend to construct and critique mechanisms over days, weeks, or months to gain traction on a goal that matters to them. A particular mechanism, and how/whether it maps onto what a professional scientist might construct, is not "the point" of such reasoning. Flint community members continued to refine their mechanisms until meaningful action was taken.

Although our argument has focused chiefly on how a community reasoned mechanistically, we acknowledge that a person may also employ mechanistic reasoning to address more-or-less individual questions or concerns. For example, an espresso machine technician might construct, and critique causal mechanisms related to scale buildup in solenoid valves to diagnose and repair a broken machine. There are likely to be important differences in what constitutes productive individual and community mechanistic reasoning, and how school might support such reasoning. Unpacking these in-detail is beyond the scope of this chapter. However, whether we aim to support individuals in making traction on their goals or communities in affecting policy change, our work should be informed by empirical accounts of people using mechanisms in real-world contexts. We should not assume that connecting science facts in an "expert-like" way to please an authority will somehow benefit individuals or society.

So how might we make formal learning spaces that enable reasoning such as what we saw among Flint residents? While acknowledging there are no simple answers to this question, here we put forward three design principles synthesized from relevant literature. Learning spaces with the potential to support mechanistic reasoning useful in-life should: (1) engage students in work rooted in the concerns/interests of their communities, (2) open space for students to grapple with complex, uncertain problems to which no obvious "right answer" exists, and (3) communicate the potential value of a variety of ways of knowing and learning. We briefly sketch each of these design principles below:

Our first design principle follows from the argument that students should engage in mechanistic reasoning for a purpose that is sensible and/or useful to them and their communities. That is, the motivation to construct mechanistic accounts cannot simply be, "I want to better understand this science concept because I (or more likely, my teacher) think it is interesting in its own right." While this sentiment might seem benign, many scholars have argued that decontextualized science learning upholds the Western modernity narrative that scientific knowledge exists outside of time, place, and culture (Bang & Medin, 2010; Popkewitz, 2022), and that all science represents progress (Morales-Doyle, 2024; Sanchez, 2023). Teaching students to view science as objective, universal, and neutral directly diminishes our capacity to develop socially, civically, or environmentally responsible students. Instead, Sanchez (2024) proposes that we connect science learning to "matters of consequentiality—sociopolitical and socio-ecological concerns facing and impacting students' communities, pasts, presents, and futurities" (p. 130). Morales-Doyle (2024) recommends that we teach science as a "catalyst for alternative futures" by orienting learning around social justice science issues (p. 149). While it is no small task to identify issues that are consequential to our students, or to center curriculum around these issues, we believe this is key to making mechanistic reasoning useful and valuable outside of a school setting.

Our second design principle is derived from the observation that mechanistic reasoning is often most useful when grappling with complex, uncertain scenarios where there is no obvious or known "right answer." For example, LeeAnn and other residents of Flint built and refined mechanistic accounts, in part, to manage "complexity, confusion, or doubt" (Dewey, 1910, p. 12; Engle, 2011) encountered when considering (mis)alignment between official accounts and lived experiences. Doing so let this community focus attention on the activities of specific entities in order to understand their experience and advocate for infrastructural changes. If we want classes to see mechanistic reasoning as a useful way of managing uncertainty, then class activities need to be structured around making sense of messy, uncertain phenomena, such as the adverse health effects experienced by the people of Flint. It is likely that classes will not have the time or expertise to arrive at an elaborated, canonical explanation for such phenomena. However, so long as we are focused on the utility of mechanistic reasoning itself and not fixated on "right answers," incomplete/partially working mechanisms are not a problem. Indeed, one could persuasively argue that, to the extent we use mechanistic reasoning in daily life at all, we are often working off of (over)simplified mechanisms that domain experts would take issue with. This is fine for most people most of the time.

Our third and final design principle is an attempt to foreground epistemic justice (Fricker, 2007) in how we imagine useful in-class mechanistic reasoning. Specifically, we claim that epistemically just engagement in mechanistic reasoning requires that classes adopt an expansive perspective on what and whose knowledge counts (Bang & Medin, 2010; Bang & Vossoughi, 2016). In practice, this means that learners should experience authority to negotiate criteria for useful mechanisms, including what knowledge should inform mechanistic accounts. Such a class would not impose a canonically correct knowledge product as the only or best mechanism. Indeed, there may be occasions in which mechanisms that diverge from a Western scientific understandings are experienced as useful by the class. For example, one could imagine a useful mechanism in which the class worked to understand how and why the Flint river became sick and what might be done to make this waterway well. Past scholarship has shown that de-settling Western perspectives on life and non-life can open new and powerful approaches to meaning-making (Bang et al., 2012). As such, exploring what makes a river unwell may be a generative approach to constructing mechanisms that support civic action. An epistemically just classroom would celebrate a mechanism of this sort and treat Indigenous knowledge (as well as other knowledge traditions) as powerful and valid.

Overall, these three principles are designed such that content and canonical ways of reasoning should not be experienced by the class as goals unto themselves. Instead, specific knowledge and ways of reasoning should be experienced as potential tools (Russ & Berland, 2019) that can help communities understand and possibly take action on issues that affect them.

In closing, we would like to acknowledge that supporting mechanistic reasoning that has potential utility in-life would require major structural changes to school and schooling. Collectively making sense of a local issue to advance more just social conditions is not compatible with systems designed to rank individual students according to who mimics "experts" with highest fidelity. Curricular, assessment, and interactional messages about what and whose knowledge counts (Russ, 2018) would have to shift from "knowledge is justified by aligning with an authoritative source" and "class is the only useful source of knowledge" toward messages like "knowledge is justified by peer consensus" and "knowledge from a variety of sources, including lived experience, is useful here." Much theoretical and empirical work remains to be done before we can claim we are supporting mechanistic reasoning that is useful in daily life.

Acknowledgements We gratefully acknowledge funding from the National Science Foundation for RLS and LBW (DRL 2003680). Any opinions, findings, and conclusions or recommendations expressed in this chapter are those of the authors and do not necessarily reflect the views of the National Science Foundation.

References

Ahn, W., & Kalish, C. W. (2000). The role of mechanism beliefs in causal reasoning. In F. Keil & A. Wilson (Eds.), *Explanation and cognition* (pp. 199–226). MIT Press.

Bachtiar, R. W., Meulenbroeks, R. F. G., & van Joolingen, W. R. (2021). Stimulating mechanistic reasoning in physics using student-constructed stop-motion animations. *Journal of Science Education and Technology, 30*(6), 777–790. https://doi.org/10.1007/s10956-021-09918-z

Bang, M., & Medin, D. (2010). Cultural processes in science education: Supporting the navigation of multiple epistemologies. *Science Education, 94*(6), 1008–1026. https://doi.org/10.1002/sce.20392

Bang, M., Warren, B., Rosebery, A. S., & Medin, D. (2012). Desettling expectations in science education. *Human Development, 55*(5–6), 302–318. https://doi.org/10.1159/000345322

Buckley, P., & Fahrenkrug, E. (2020). The Flint, Michigan water crisis as a case study to introduce concepts of equity and power into an analytical chemistry curriculum. *Journal of Chemical Education, 97*(5), 1327–1335. https://doi.org/10.1021/acs.jchemed.9b00669

City of Flint. (2015). *Important information about your drinking water: City of Flint did not meet treatment requirements*. https://www.cityofflint.com/wp-content/uploads/TTHM-Notification-Final.pdf

Clark, A. (2018). *The poisoned city: Flint's water and the American urban tragedy* (First). Metropolitan Books. https://us.macmillan.com/books/9781250125149/thepoisonedcity

Cooper, M., & Klymkowsky, M. (2013). Chemistry, life, the universe, and everything: A new approach to general chemistry, and a model for curriculum reform. *Journal of Chemical Education, 90*(9), 1116–1122. https://doi.org/10.1021/ed300456y

Cooper, M. M., Kouyoumdjian, H., & Underwood, S. M. (2016). Investigating students' reasoning about acid–base reactions. *Journal of Chemical Education, 93*(10), 1703–1712. https://doi.org/10.1021/acs.jchemed.6b00417

Dewey, J. (1910). *How we think*. Heath.

Dingle, A. (2016, December). The Flint water crisis: What's really going on? *Chemmatters*, 1–8.

Dood, A. J., Dood, J. C., Cruz-Ramírez de Arellano, D., Fields, K. B., & Raker, J. R. (2020). Analyzing explanations of substitution reactions using lexical analysis and logistic regression techniques. *Chemistry Education Research and Practice, 21*(1), 267–286. https://doi.org/10.1039/C9RP00148D

Edwards, M., Roy, S., & Rhoads, W. (2015, September). *Lead testing results for water sampled by residents*. Flint Water Study Updates: Up-to-Date Information on Our Collaborative Research and Citizen Science Work with the Residents of Flint, MI in Light of Reported Water Quality Issues.

Engle, R. A. (2011). The productive disciplinary engagement framework: Origins, key concepts, and developments. In D. Dai (Ed.), *Design research on learning and thinking in educational settings: Enhancing growth and functioning*. Routledge.

Feinstein, N. W., & Baram-Tsabari, A. (2024). Epistemic networks and the social nature of public engagement with science. *Journal of Research in Science Teaching*. https://doi.org/10.1002/tea.21941

Fleming, L. N. (2018, February 5). Expert: Flint plant not ready before water switch. *The Detroit News*. https://www.detroitnews.com/story/news/michigan/flint-water-crisis/2018/02/05/flint-hearing-plant-unready-deq/110124346/#

Flint Water Advisory Task Force. (2016). *Final Report*. https://www.michigan.gov/-/media/Project/Websites/formergovernors/Folder6/FWATF_FINAL_REPORT_21March2016.pdf?rev=284b9e42c7c840019109eb73aaeedb68

Fricker, M. (2007). *Epistemic injustice: Power and the ethics of knowing*. Clarendon Press.

Glennan, S. (2002). Rethinking mechanistic explanation. *Philosophy of Science, 69*(S3), S342–S353. Cambridge Core. https://doi.org/10.1086/341857

Koslowski, B. (1996). *Theory and evidence: The development of scientific reasoning*. MIT Press.

Machamer, P., Darden, L., & Craver, C. F. (2000). Thinking about mechanisms. *Philosophy of Science, 67*(1), 1–25. https://doi.org/10.1086/392759

Miller, E., Manz, E., Russ, R., Stroupe, D., & Berland, L. (2018). Addressing the epistemic elephant in the room: Epistemic agency and the next generation science standards. *Journal of Research in Science Teaching, 55*(7), 1053–1075. https://doi.org/10.1002/tea.21459

Morales-Doyle, D. (2024). *Transformative science teaching: A catalyst for justice and sustainability*. Harvard Education Press.

National Research Council. (2013). *Developing assessments for the next generation science standards*. National Academies Press. https://www.nap.edu/catalog/18409/developing-assessments-for-the-next-generation-science-standards

NGSS Lead States. (2013). *Next generation science standards: For states, by states*. The National Academies Press.

Passmore, C., Gouvea, J. S., Giere, R., & Matthews, M. R. (2014). Models in science and in learning science: Focusing scientific practice on sense-making. In *International handbook of research in history, philosophy and science teaching*. Springer.

Popkewitz, T. S. (2022). Making the scientific self: A location-less logic with locations. In *Epistemic colonialism and the transfer of curriculum knowledge across borders* (pp. 21–44). Routledge.

Rudolph, J. L. (2023). *Why we teach science: (and why we should)*. Oxford University Press. https://books.google.com/books?id=1J5JzwEACAAJ

Russ, R. S. (2014). Epistemology of science vs. epistemology for science. *Science Education, 98*(3), 388–396. https://doi.org/10.1002/sce.21106

Russ, R. S. (2018). Characterizing teacher attention to student thinking: A role for epistemological messages. *Journal of Research in Science Teaching, 55*(1), 94–120. https://doi.org/10.1002/tea.21414

Russ, R. S., Coffey, J. E., Hammer, D., & Hutchison, P. (2009). Making classroom assessment more accountable to scientific reasoning: A case for attending to mechanistic thinking. *Science Education, 93*(5), 875–891. https://doi.org/10.1002/sce.20320

Russ, R. S., Scherr, R. E., Hammer, D., & Mikeska, J. (2008). Recognizing mechanistic reasoning in student scientific inquiry: A framework for discourse analysis developed from philosophy of science. *Science Education, 92*(3), 499–525. https://doi.org/10.1002/sce.20264

Russ, R. S., & Berland, L. K. (2019). Invented science: A framework for discussing a persistent problem of practice. *Journal of the Learning Sciences, 28*(3), 279–301. https://doi.org/10.1080/10508406.2018.1517354

Sanchez, A. (2023). The social focus framework: Antiracist and anticolonial conscientization, consequence, and presencing in science education. In *Reimagining science education in the anthropocene* (Vol. 2, pp. 121–142). Springer International Publishing.

Schwarz, C. V., Passmore, C., & Reiser, B. J. (2016). Helping students make sense of the world using next generation science and engineering practices—PB351X. National Science Teachers Association – NSTA Press.

Schwarz, C. V., Reiser, B. J., Davis, E. A., Kenyon, L., Achér, A., Fortus, D., Shwartz, Y., Hug, B., & Krajcik, J. (2009). Developing a learning progression for scientific modeling: Making scientific modeling accessible and meaningful for learners. *Journal of Research in Science Teaching, 46*(6), 632–654. https://doi.org/10.1002/tea.20311

Shwartz, Y., Weizman, A., Fortus, D., Krajcik, J., & Reiser, B. (2008). The IQWST experience: Using coherence as a design principle for a middle school science curriculum. *The Elementary School Journal, 109*(2), 199–219. https://doi.org/10.1086/590526

Terry, T. J. (2017, December 5). *The flint water crisis: An introduction to chemical reactions*. NSTA: NCCSTS Case Collection. https://www.nsta.org/ncss-case-study/flint-water-crisis

The National Research Council. (2012). *A framework for K-12 science education: Practices, crosscutting concepts, and core ideas*. National Academies Press. http://nap.edu/catalog/13165

Torrice, M. (2016). How lead ended up in flint's water. *Chemical and Engineering News, 94*(7), 26–29.

Westfall, R. S. (1986). *The construction of modern science: Mechanisms and mechanics*. Cambridge University Press.

Yu, A., & Linden, L. E. (2022). Environmental and social injustices of the flint, MI water crisis: A general chemistry exercise. *Journal of Chemical Education, 99*(1), 469–471. https://doi.org/10.1021/acs.jchemed.1c00371

Chapter 13
Mechanistic Literacy Across Functional Health-Related Daily Life Decisions

Ilana Dubovi

13.1 Introduction

One of the primary objectives of K–12 science education is to equip students with the knowledge and skills necessary to thrive in a scientifically and technologically advanced society (Feinstein, 2011; National Academies of Sciences, Engineering, and Medicine, 2016; Sharon & Baram-Tsabari, 2020; Tabak, 2018). In particular, science education seeks to cultivate individuals as "competent outsiders" (Feinstein, 2011) who are able to recognize when and how science is relevant to their functional needs and interests in their everyday civic life (Ryder, 2001). Importantly, competent outsiders are individuals who, by connecting scientific knowledge with everyday life, are able to make informed decisions and enrich their understanding of their personal circumstances (Feinstein, 2011; Feinstein et al., 2013). Numerous recent studies have examined scientific practices used by laypersons in everyday life, with the aim of informing the design of educational experiences and curricula on how to prepare students to reason with and about scientific evidence to become "competent outsiders." However, these studies focus mainly on data literacy (Lee & Dubovi, 2020; Tabak & Dubovi, 2023), information-seeking (Shauli & Baram-Tsabari, 2019), and real-world tasks to reveal laypersons' scientific practices (Bromme et al., 2015; Gottschling et al., 2020; Stadtler & Bromme, 2004), with less attention given to the perspective of mechanistic literacy.

To fill this gap and reveal how laypersons use mechanistic accounts in their everyday lives, in this chapter, we examine the functional context of health-related issues by focusing on two case studies that address how patients deal with health

I. Dubovi (✉)
Department of Nursing Sciences, Gray Faculty of Medical & Health Sciences, Tel Aviv University, Tel Aviv-Yafo, Israel
e-mail: ilanadubovi@tauex.tau.ac.il

N. Graulich et al. (eds.), *Exploring Mechanistic Reasoning in Science Education*, Contributions from Science Education Research 17,
https://doi.org/10.1007/978-3-032-10246-1_13

issues. Specifically, we investigate patients who are facing gastrointestinal (GI) cancer or type 1 diabetes mellitus (T1DM), and are therefore inherently compelled to make daily disease-related decisions. This presents a unique opportunity to better understand how "forced choice" daily life circumstances and bodily "episodic feelings" when facing a disease can trigger, sustain, and support scientific practices through the lens of mechanistic reasoning (Nemirovsky, 2011).

13.2 Mechanistic Literacy

Mechanistic literacy refers to the understanding of the mechanisms underlying scientific phenomena. This kind of literacy allows individuals to comprehend *how* the activities of the constituent system components interact to produce the overall effect of a phenomenon (Bolger et al., 2012). Koslowski (1996) proposed that mechanistic accounts involve "the process by which a cause brings about an effect" (p. 13). Russ et al. (2008) and (Hammer, 1995) distinguished between causal inferences based on input–output relations—the "causal intuitions" that emerge very early during human development, and the mechanistic conception of specific processes within particular domains. In contrast to causal inferences, mechanistic accounts include "the process by which a cause" brings about an effect, including entities' properties and activities at at least one scale level below the scale level of a target phenomenon (Dickes et al., 2016; Koslowski, 1996; Krist et al., 2019; Wilensky & Resnick, 1999). For instance, while within the context of cancer treatment, understanding that chemotherapy might cause multiple side effects would be a cause-and-effect inference, mechanistic understanding would emphasize how chemotherapy targets rapidly dividing cells of both cancerous and certain healthy tissues (Haskel-Ittah & George-Levi, 2023). Similarly, regarding T1DM, reasoning about the effect of insulin injections on reducing blood glucose levels would be a cause-and-effect inference. A more mechanistic account of how insulin mediates glucose uptake into muscle cells, which then undergo glycolysis to produce energy as ATP, would be a deep explanation that identifies both the causes and the underlying processes (Dagan et al., 2019).

Such mechanistic explanations have been suggested to be essential for developing scientific literacy (Russ et al., 2008) and thus have been chosen as one of the crosscutting concepts in the Framework for K–12 Science Education (National Research Council, 2012) and the Next Generation Science Standards (NGSS Lead States, 2013). However, because most of the research is focused on the professional context—how a scientist thinks, and less on functional laypersons' mechanistic use of mechanistic accounts in their daily lives, little is known about how to cultivate mechanistic literacy. Moreover, research that does exist on mechanistic literacy among the public has introduced inconsistent findings. Whereas Dagan et al. (2019) showed that an understanding of causal biochemical processes related to glucose equilibrium is associated with better decisions related to diabetes management, Taube et al. (2021) showed that the provision of mechanistic information does not increase acceptance of the notion that climate change is anthropogenic. These mixed findings highlight the need for further studies focused on how and whether mechanistic literacy is relevant to the daily needs and interests of laypersons.

13.3 The Context of Health-Related Daily Life

In this chapter, we focus on two health-related functional contexts of everyday life: the experiences of cancer patients undergoing chemotherapy and those of patients managing T1DM. Both cancer and T1DM management require active participation in daily self-care decisions. Patients who have been diagnosed with cancer and are undergoing chemotherapy face, on average, 11 to 13 different side effects simultaneously every day, such as peripheral neuropathy, nausea, pain, and mucositis (Lin et al., 2020). Given the trend away from hospitalization and toward outpatient care, patients with cancer often experience those adverse effects at home; therefore, they are expected to learn how to monitor, manage, and prioritize side effects by themselves, with no healthcare professional by their side (Bubis et al., 2018; Corsi et al., 2020). For example, chemotherapy might cause neutropenia, a severe drop in white blood cells, making patients highly susceptible to infections. If a patient develops any mild symptoms of infection, such as a low-grade fever, a sore throat, extreme fatigue, or a runny nose, they need to recognize the urgency and seek immediate medical attention. Without it, the infection could lead to sepsis, a life-threatening condition. Another example of daily decision-making regarding chemotherapy side effects relates to the prevention of nausea and vomiting. Without adhering to a strict regimen of various medications, such as taking promethazine initially and ondansetron (Zofran) after 3 days, severe dehydration can occur. This can lead to weakness and dizziness and disrupt the balance of essential electrolytes. Severe dehydration can cause kidney failure and other critical issues.

Such active patient participation in disease self-management is also relevant for patients who have been diagnosed with T1DM and, as a result, need to monitor and maintain their blood glucose levels while receiving insulin therapy. T1DM is an autoimmune disease that requires lifelong insulin replacement to maintain a delicate balance of blood glucose levels. Therefore, T1DM patients manage the risks of hypoglycemia and hyperglycemia daily by performing invasive procedures (blood glucose monitoring and insulin injections) and maintaining a strict treatment and nutrition regimen. For example, during a morning jog, a T1DM patient might suddenly feel dizzy and weak. Very quickly, they need to recognize that this could be due to low blood glucose levels. Since hypoglycemia is potentially life-threatening, it is essential for them to understand the importance of not only carrying a finger-prick test kit but also making the decision to perform the test immediately. In addition, they should have fast-acting carbohydrates, such as glucose tablets or juice, on hand to quickly raise their blood sugar level if needed. Another example of daily decision-making routines relates to the continuous need to count the carbohydrates in their meals. If a patient miscalculates the insulin dosage, they might suffer from hypoglycemia or hyperglycemia.

Active patient participation in disease self-management is functional not only due to the ongoing shift in healthcare from inpatient to outpatient care, but also due to the shift of medical care from the patriarchal patient–clinician relationship to a model of shared decision-making. In contrast to the traditional patient–clinician relationship, where the clinician makes decisions on behalf of the patient, in shared decision-making, the patient and healthcare professionals collaborate to reach a treatment decision informed by evidence and the patient's goals of care (Joosten et al., 2008; Punnett et al., 2024). This approach recognizes the importance of patients' active involvement in their own healthcare decisions, ensuring that their preferences, values, and needs are considered alongside clinical evidence (Barry & Edgman-Levitan, 2012). While previous research in science education acknowledges that laypersons differ from experts in their ability to comprehend and evaluate evidence (Barzilai & Chinn, 2020; Duncan et al., 2018), healthcare systems often encourage patients to narrow this expert–novice gap. This is especially true when a second expert opinion is needed to make an informed choice between different medical options, or when there are various treatment options to choose from. Therefore, since the goal of cultivating a science-literate public fits in with the healthcare system policy, it establishes the framework for the current study on mechanistic accounts in functional scientific reasoning.

13.4 The Present Study

In this chapter, we present two case studies that are part of a larger ongoing study aimed at revealing how laypersons use science in everyday life (Dubovi et al., 2024); our ultimate aim is to inform the design of science education experiences that foster functional scientific literacy (Feinstein, 2011). In the context of health-related daily life issues, we ask: (1) Does the public use mechanistic accounts in functional scientific reasoning? (2) If so, what characterizes their functional mechanistic reasoning? (3) How important is mechanistic literacy for functional daily life? To answer these questions, we first introduce the current study's participants, followed by the research tools and analysis.

13.4.1 Participants

Participants in this study were patients diagnosed with health conditions. We recruited 72 adult patients from the cancer unit in a tertiary hospital who were diagnosed with the most common GI cancer diseases (i.e., gastric, liver, colorectal, esophagus, and pancreas) and were being treated with chemotherapy. Those cancer patients were 63.4 ± 12.3 years of age on average, and 29 (40%) were female. GI cancer disease diagnosis duration was, on average, 1.54 ± 0.9 years. We also recruited 54 adult patients diagnosed with T1DM at a diabetes clinic in a tertiary hospital who were being treated with insulin via multiple daily injections or continuous subcutaneous insulin infusion. On average, participants with T1DM were 52.6 ± 16.9 years of age, and 27 (50%) were female. The disease diagnosis was made at 25.1 ± 16.5 years, and its duration averaged 27.1 ± 14.1 years.

A comparison of sociodemographic characteristics revealed that the cancer patients were significantly older than the T1DM patients ($t = 3.78$, $p < 0.001$). However, the T1DM patients had been managing their health condition from diagnosis for a significantly longer period, on average 17 times longer than cancer patients ($t = -13.02$, $p < 0.001$).

13.4.2 Measures and Findings

13.4.2.1 Mechanistic Accounts

To evaluate whether and how the public uses mechanistic accounts to make sense of their own everyday lives in the context of health, we used a questionnaire that included an open-ended question relevant to patients' disease issues. We analyzed patients' explanations using two approaches: thematic analysis and a causal reasoning coding system to examine the connections between elements in their explanations. Here, we present the open-ended question and its analysis, first for oncology patients, and then for T1DM patients.

Oncology patients were posed an open-ended question, previously developed by the author (Dubovi et al., 2024), which asked them to explain why chemotherapy treatment causes specific side effects, such as hair loss, a weakened immune system, mouth sores, and more. Thematic analysis identified eight recurring elements in the patients' explanations: (1) damage to both healthy and cancerous cells (18 occurrences); for example, "the chemical agents in chemotherapy destroy all cells, both harmful and healthy"; (2) side effects as a byproduct of treatment (15 occurrences); for example, "the fatigue and hair loss are side effects caused by the treatment aimed at eradicating cancer but also result in additional harm"; (3) immune system suppression (14 occurrences); for example, "chemotherapy damages the immune system, making the body more vulnerable to life-threatening illnesses"; (4) rapidly dividing cells (10 occurrences); for example, "the treatment targets cells that divide rapidly, such as cancer cells, but also affects other rapidly dividing cells in the body"; (5) systemic effects and non-specificity (9 occurrences); for example, "the chemo treatment impacts the entire body, not just the tumor, resulting in multiple side effects"; (6) chemical nature of chemotherapy (8 occurrences); for example, "chemotherapy is a type of chemical toxin designed to kill cancer cells, but it also affects healthy cells"; (7) targeting specific cells (7 occurrences); for example, "the treatment is intended to target cancer cells specifically but also negatively impacts other systems in the body"; (8) regeneration and recovery of healthy cells (6 occurrences); for example, "healthy cells have the ability to regenerate themselves after treatment ends, unlike cancer cells."

In addition, a bottom-up approach, based on Haskel-Ittah and George-Levi's (2023) causal reasoning coding system, was used to analyze the connections between elements in patients' explanations. The following four categories of explanations were identified: (1) patients who stated that they did not know how to explain the phenomena (35%); (2) patients who attributed chemotherapy side effects to the inherent strength and perceived harmful nature of chemotherapy medications (27%), stating that the therapy is "strong" or "poisonous"; (3) patients who used causal inferences without mechanistic grounds (5%). The category of causal inferences included two parallel causal channels: (i) chemotherapy has an effect on the tumor, (ii) chemotherapy impacts healthy tissue and thus causes side effects (Fig. 13.1a). For example, one of the patients explained: "The chemotherapy treatment involves many strong drugs that are supposed to target the cancer, but along the way, they also affect various systems in the body"; (4) patients who used causal inferences that incorporated mechanistic components to explain why and how chemotherapy causes specific side effects (Fig. 13.1b, 33%). Namely, patients pointed out the micro-level nature of rapidly dividing cells as a critical component in explaining why chemotherapy can affect both cancerous and healthy tissues. For example, one of the patients explained: "The chemotherapy treatment is aimed at destroying rapidly dividing cells; that's why it is not selective. Therefore, it also kills healthy cells, particularly those that divide quickly." Another patient elaborated: "The treatment affects cells that divide rapidly. That's why it affects the white blood cells and other cells (for example, platelets) produced in the bone marrow that are part of the immune system. Because the cells of the hair roots divide rapidly, the treatment also affects them."

Diabetes patients were asked to answer the following open-ended question, developed by the author previously (Dagan et al., 2019; Dubovi et al., 2020): "John has T1DM. He is an athlete and is about to start a swimming race. His blood tests show very high sugar levels (385 mg%). Please describe the impact of such high blood glucose on John and his chances of winning the competition." This question aimed to reveal the reasoning that connects glucose to cellular energy production, with insulin acting as a mediator. Thematic analysis identified five recurring elements in the patients' explanations: (1) glucose (33 occurrences) was commonly emphasized; for example, "glucose is the main source of energy for the body's

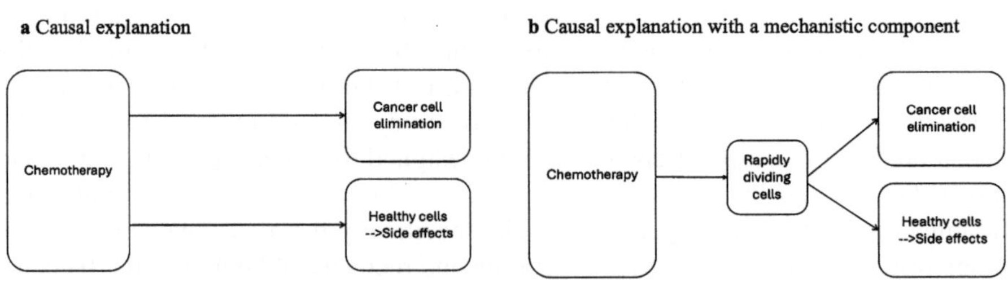

Fig. 13.1 (**a**) Causal explanation with two parallel causal channels, (**b**) Causal explanation that includes the interconnected mechanistic component of rapidly dividing cells

cells"; (2) energy production (25 occurrences), with many noting that "cells cannot produce energy without glucose entering the cell"; (3) insulin function (22 occurrences), where participants often described the role of insulin in facilitating glucose absorption; for example, "insulin allows cells to absorb glucose and use it to produce energy"; (4) high blood sugar (21 occurrences), where discussions highlighted its implications; for example, "high blood sugar levels indicate that glucose is not entering the cells"; (5) health impacts (19 occurrences); for example, "high blood sugar can cause damage to the kidneys and blood vessels."

In addition, a bottom-up qualitative analysis utilizing a causal reasoning coding scheme from previous studies (Dagan et al., 2019; Dubovi et al., 2020) revealed four distinct categories of explanations, demonstrating connections between the elements in the patients' explanations (Dagan et al., 2019; Dubovi et al., 2020): (1) 35% of patients stated that they did not know how to explain the impact of high blood glucose levels; (2) 15% explained that high glucose is not healthy, stating that "high blood glucose actually damages the kidneys and delicate blood vessels" and sharing from their own experience: "after a rise in blood sugar levels, one feels fatigued and dysfunctional"; (3) 15% of patients used causal explanations by identifying a causal link between glucose and energy production (Fig. 13.2a): "the sugar remains in the blood and does not enter the cells, and that's why they are not producing energy"; (4) 35% of patients, in addition to causal inferences between glucose and cellular energy production, identified a mechanistic component of insulin as the mediator responsible for glucose uptake from the bloodstream into the cells (Fig. 13.2b). For example, one of the patients stated: "as long as there is not enough insulin to allow sugar to enter the cells, the sugars in the blood will not be utilized for the benefit of the cells to produce energy." Another patient stated: "John's blood sugar level indicates that his beta cells are not secreting enough insulin to lower the blood sugar level. As a result, his cells are unable to absorb sugar and convert it into energy."

13.4.2.2 Involvement in Decision-Making

To assess patients' involvement and confidence in decision-making processes with their clinicians, we asked them to complete the Decision Self-Efficacy/Confidence Scale (O'Connor, 1995). This questionnaire includes 11 items rated on a Likert scale from 0 (not confident at all) to 4 (very confident). Example items include: "I

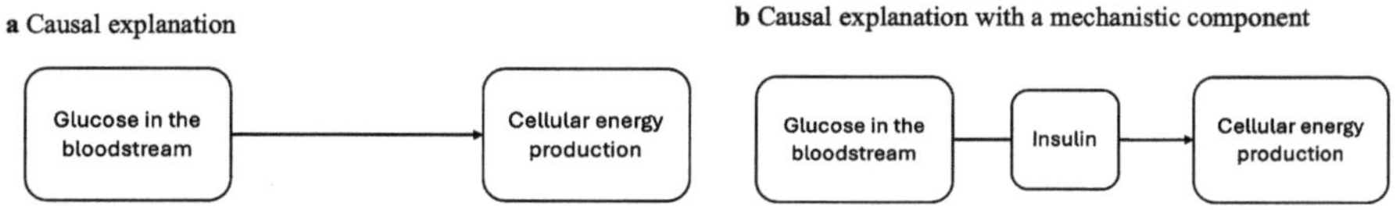

Fig. 13.2 (**a**) Causal explanation that connects glucose to cellular energy production, (**b**) Causal explanation that includes the mechanistic component of insulin, which is responsible for facilitating the uptake of glucose into cells, thereby enabling cellular energy production

feel confident that I can express my concerns about each choice," "I feel confident getting the facts about the medication choices available to me," and "I feel confident letting the clinical team know what's best for me." The reliability of the previously translated Hebrew version was $\alpha = .93$, and in the current study, it was $\alpha = .91$.

The patients diagnosed with cancer reported an average decision self-efficacy level of 3.4 ± 0.9, whereas patients with T1DM reported significantly higher levels of decision self-efficacy, 3.7 ± 0.4 ($t = -2.202$, $p < 0.05$). To evaluate the associations between self-efficacy and the level of involvement and confidence in decision-making, we used a bivariate Spearman's rank correlation analysis, suitable for our ordinal data. The analysis demonstrated a significant association between the tendency to incorporate causal inferences and decision self-efficacy for cancer patients ($r_{s=}$ 0.265, $p < 0.05$), as well as for T1DM patients ($r_{s=}$ 0.361, $p < 0.05$).

13.5 Discussion

Many studies have focused on how to better prepare learners for civic participation by emphasizing data literacy (Lee & Dubovi, 2020), information-seeking (Shauli & Baram-Tsabari, 2019), and real-world tasks to reveal laypersons' scientific practices (Bromme et al., 2015; Gottschling et al., 2020). The contribution of the current study lies in its focus on mechanistic literacy. To address this, we explored the functional scientific practices of cancer patients and T1DM patients, focusing on their disease management. Building upon Feinstein's (2011) concept of "competent outsiders," describing non-scientists who are capable of accessing and making "meaningful" use of science to solve problems and make decisions on issues that matter to them in their own lives, this study acknowledges that what qualifies as "meaningful" depends on the context. We show that the "meaningful" use of science becomes especially important in the context of disease self-care due to its almost immediate impact on personal life, sometimes involving matters of life or death. Moreover, disease management provides a fertile context for the current inquiry because healthcare system policies encourage patients to not merely place their trust in clinician experts (Hendriks & Kienhues, 2019), but rather gain a certain level of medical competency to handle shared decision-making with their physician. Using functional questions to explore GI cancer and T1DM as focal cases of laypersons' scientific literacy yielded three main takeaways that address current research questions. In the following, we elaborate on each of them.

13.5.1 First Takeaway—Public Tendency to Incorporate Mechanistic Accounts

The findings showed that some laypersons use mechanistic reasoning to make sense of their own disease. Although we investigated two very different health conditions—one being a chronic condition that patients have managed since childhood (namely, T1DM) and the other a relatively new diagnosis but an extreme case involving life or death decisions (namely, GI cancer), we found that the tendency to incorporate mechanistic components in causal reasoning was relatively similar across both groups. In particular, about 33% of patients diagnosed with cancer and about 30% of patients diagnosed with T1DM incorporated mechanistic biological accounts as part of their reasoning. Whereas recent discussions concerning the "post-truth" era present people as disinclined to draw on scientific practices (Barzilai & Chinn, 2020; Valladares, 2021), the current study portrays a more optimistic perspective, demonstrating that laypersons can incorporate scientific practices in terms of mechanistic reasoning. The propensity of laypersons to use science as part of their functional everyday lives has also been demonstrated previously in the context of data literacy. For example, Golumbic et al. (2020), as part of the citizen science project, evaluated laypersons' ability to understand and interpret air quality data presented in various visual formats. The results showed that 70% of the lay participants were able to interpret the data presented in various visual representations. Another study by Tabak and Dubovi (2023), investigating laypersons' data literacy and tendencies during the COVID-19 pandemic, showed that about 55% of participants incorporated data into their decision-making processes.

13.5.2 Second Takeaway—Characterization of Functional Mechanistic Reasoning

Although some patients did use mechanistic accounts in their explanations, they were incomplete, lacking many details on related components, their activities, and properties; they often integrated causal inferences with only a single mechanistic component. For example, cancer patients attributed the causal relationship between chemotherapy and side effects to the underlying mechanism of targeting rapidly dividing cells. However, we assume that an expert would have provided a much more detailed explanation, including mechanisms of drug action related to cell-cycle phases leading to activation of cell apoptosis through regulatory proteins such as p53, and to cell death through various mechanisms, such as cross-linking of DNA strands, inhibition of enzymes essential for DNA replication, disruption of microtubule function, or interference with cell signaling pathways critical for cell survival. The T1DM patients' explanations were also incomplete. They connected the causal pathway between glucose and cellular energy production through the insulin mechanism. However, many of the mechanistic details were missing, such as types of cells (muscle cells versus beta cells), GLUT-4 transporter, and the multistep process of glucose conversion to ATP, involving glycolysis and the Krebs cycle.

Similar incomplete mechanistic reasoning was also demonstrated in studies of university students who were asked to generate explanations of genetic and chemistry-related phenomena (Scalco et al., 2018; Sevian et al., 2018; Southard et al., 2017). In those studies, the university students could only recall one mechanistic component. The discrepancy between a possible detailed expert mechanistic explanation and that of laypersons can be understood through the perspective of "explanatory black boxes," elaborated for science education by Haskel-Ittah (2023). Within the context of mechanistic reasoning, those black boxes are defined as providing a gap between one step of the mechanistic process and the next (Arie Livni & Michal Haskel-Ittah, current book). Cognitive science shows that unpacking everything related to a phenomenon and its activities is not feasible for non-experts, as they cannot mentally grasp and retain the complete details of a complex mechanism (Keil et al., 2008; Trouche et al., 2017). Moreover, as we are all non-experts in the context of most daily life issues (Osborne & Pimentel, 2023), detailed mechanistic reasoning might not function to support decision-making in everyday life. Instead, incomplete mechanistic inferences, where irrelevant details are dropped to serve a functional purpose, can be sufficient to activate causal links and a sense of causal patterns to support decisions (Craver & Kaplan, 2020). To this end, we suggest that laypersons' incomplete mechanistic reasoning represents the *functional mechanistic reasoning of competent outsiders.*

13.5.3 Third Takeaway—The Importance of Mechanistic Literacy for Functioning in Daily Life

This takeaway supports the notion of the importance of mechanistic explanations for functioning in daily life. Our results show that patients' tendency to incorporate mechanistic explanations, even if they are incomplete, is associated with a higher-level perception of self-efficacy to engage in an active decision-making process. Because engagement in decision-making has been clearly demonstrated to improve health outcomes, this shows that mechanistic literacy is important for functioning in daily life (Montori et al., 2006; Tamhane et al., 2015; Vo et al., 2023).

The importance of mechanistic reasoning in decision-making can also be illustrated through the following examples. For patients with T1DM, understanding how insulin facilitates glucose uptake into cells and its role in cellular respiration for energy production can help patients interpret symptoms of hypoglycemia—such as weakness, dizziness, and confusion—as resulting from a lack of energy production in their cells. During physical activity, such as jogging, a patient with T1DM may experience a sudden drop in blood glucose levels. Recognizing this as hypoglycemia—a potentially life-threatening condition, enables them to act quickly by consuming fast-acting carbohydrates, such as glucose tablets or juice, to restore normal glucose levels. This mechanistic understanding can reinforce patients' adherence to timely insulin injections with meals and the maintenance of basal insulin levels to ensure continuous cellular energy production.

Moreover, patients often receive a long list of guidelines, which can be overwhelming and difficult to remember. A deep mechanistic understanding enables patients to prioritize and flexibly apply those guidelines in novel or unusual situations. For example, in cases of alcohol consumption, a patient who understands the insulin–glucose mechanism would recognize the heightened risk of hypoglycemia. Alcohol can interfere with the liver's ability to release glucose into the bloodstream from liver cells, making it critical to avoid unnecessary insulin injections during or after drinking, even if carbohydrates have been consumed. Similarly, during a fever, patients might be advised to maintain or even increase their insulin dosage, even if they are not eating. This is because fever triggers the release of stress hormones, such as cortisol and adrenaline, which increase blood glucose levels. A mechanistic understanding would support the patient's grasp of why insulin is still crucial in this scenario—to counteract the hyperglycemic effects of stress hormones and prevent complications such as diabetic ketoacidosis.

13.5.4 Final Words

Science education aims to equip students with the skills to critically engage in the scientific and data-driven issues that they encounter in their personal and civic lives. Analyzing two case studies of laypersons' functional reasoning, we suggest that to achieve this, students need more than just information sourcing or the use of heuristics to identify inaccurate information (Duncan et al., 2018); they also need to establish a foundation for functional mechanistic reasoning to support their informed decisions in civic daily life. Toward future work, we are looking into expanding the depth of the current case studies by developing more open-ended functional questions using a longitudinal perspective to pinpoint changes in mechanistic reasoning with time. Furthermore, to cultivate students' functional mechanistic reasoning and scientific practices, we aim to develop a curricular focus that engages students with science not as junior scientists, but as laypersons (Chinn et al., 2023).

13.5.5 Limitations

The present study has several limitations. It relies primarily on self-report measures to capture both patient reasoning and their decision-making processes. While informative, these measures may not fully represent the dynamic interactions between reasoning processes and real-time decision-making. Nevertheless, the study was conducted in naturalistic settings, with the questionnaires serving as a practical proxy for addressing patients' day-to-day functional questions. Furthermore, the use of contextualized questionnaires aligns with common practices in science education research, particularly in studies focused on laypersons' literacy and decision-making (e.g., Rozenblum et al., 2024; Tabak & Dubovi, 2023). Future studies should include additional methods, such as interviews alongside the questionnaires, and extend these to include patients' family members and caregivers, who often play a significant role in disease management and decision-making.

In addition, the study sample was relatively small and focused only on two diseases—GI cancer and T1DM. This narrow scope highlights the need for future research to explore how laypersons with other conditions (e.g., asthma, Crohn's disease) and those without chronic conditions (i.e., healthy individuals) engage in functional mechanistic reasoning. Additional studies with larger and more diverse samples are warranted to validate and expand upon these findings.

Acknowledgments The Israel National Institute for Health Policy Research (grant no. P135767).

References

Barry, M. J., & Edgman-Levitan, S. (2012). Shared decision making – The pinnacle of patient-centered care. *New England Journal of Medicine, 366*(9), 780–781). Massachusetts Medical Society. https://doi.org/10.1056/NEJMp1109283

Barzilai, S., & Chinn, C. A. (2020). A review of educational responses to the "post-truth" condition: Four lenses on "post-truth" problems. *Educational Psychologist, 55*(3), 107–119. Routledge. https://doi.org/10.1080/00461520.2020.1786388

Bolger, M. S., Kobiela, M., Weinberg, P. J., & Lehrer, R. (2012). Children's mechanistic reasoning. *Cognition and Instruction, 30*(2), 170–206. https://doi.org/10.1080/07370008.2012.661815

Bromme, R., Thomm, E., & Wolf, V. (2015). From understanding to deference: Laypersons' and medical students' views on conflicts within medicine. *International Journal of Science Education, Part B: Communication and Public Engagement, 5*(1), 68–91. https://doi.org/10.1080/21548455.2013.849017

Bubis, L. D., Davis, L., Mahar, A., Barbera, L., Li, Q., Moody, L., Karanicolas, P., Sutradhar, R., & Coburn, N. G. (2018). Symptom burden in the first year after cancer diagnosis: An analysis of patient-reported outcomes. *Journal of Clinical Oncology, 36*(11), 1103–1111. https://doi.org/10.1200/JCO.2017.76.0876

Chinn, C. A., Yoon, S. A., Hussain-Abidi, H., Hunkar, K., Noushad, N. F., Cottone, A. M., & Richman, T. (2023). Designing learning environments to promote competent lay engagement with science. *European Journal of Education, 58*(3), 407–421. https://doi.org/10.1111/ejed.12573

Corsi, M. P., Shea, K., & W Knoebel, R. (2020). Impact of transitioning inpatient chemotherapy regimens to the outpatient setting. *Journal of Oncology Pharmacy Practice, 26*(6), 1324–1330. https://doi.org/10.1177/1078155219891566

Craver, C. F., & Kaplan, D. M. (2020). Are more details better? On the norms of completeness for mechanistic explanations. *The British Journal for the Philosophy of Science, 71*(1), 287–319. https://doi.org/10.1093/Bjps/Axy015

Dagan, E., Dubovi, I., Levy, M., Zuckerman Levin, N., & Levy, S. T. (2019). Adherence to diabetes care: Knowledge of biochemical processes has a high impact on glycaemic control among adolescents with type 1 diabetes. *Journal of Advanced Nursing, 75*(11). https://doi.org/10.1111/jan.14098

Dickes, A. C., Sengupta, P., Farris, A. V., & Basu, S. (2016). Development of mechanistic reasoning and multilevel explanations of ecology in third grade using agent-based models. *Science Education, 100*(4), 734–776. https://doi.org/10.1002/sce.21217

Dubovi, I., Levy, S. T., Levy, M., Zuckerman Levin, N., & Dagan, E. (2020). Glycemic control in adolescents with type 1 diabetes: Are computerized simulations effective learning tools? *Pediatric Diabetes, 21*(2), 328–338. https://doi.org/10.1111/pedi.12974

Dubovi, I., Pelles-Avraham, S., Visel, T., Kashi, A., & Haskel-Ittah, M. (2024). The role of causal inferences in everyday reasoning: Study on oncology patients. In *Proceedings of the 18th international conference of the learning sciences – ICLS 2024* (pp. 1031–1034). https://doi.org/10.22318/icls2024.281365

Duncan, R. G., Chinn, C. A., & Barzilai, S. (2018). Grasp of evidence: Problematizing and expanding the next generation science standards' conceptualization of evidence. *Journal of Research in Science Teaching, 55*(7), 907–937. https://doi.org/10.1002/tea.21468

Feinstein, N. (2011). Salvaging science literacy. *Science Education, 95*(1), 168–185. https://doi.org/10.1002/sce.20414

Feinstein, N. W., Allen, S., & Jenkins, E. (2013). Outside the pipeline: Reimagining science education for nonscientists. *Science, 340*(6130), 314–317. https://doi.org/10.1126/science.1230855

Golumbic, Y., Fishbain, B., & Baram-Tsabari, A. (2020). Science literacy in action: Understanding scientific data presented in a citizen science platform by non-expert adults. *International Journal of Science Education, Part B, 10*(3), 232–247. https://doi.org/10.1080/21548455.2020.1769877. Taylor & Francis.

Gottschling, S., Kammerer, Y., Thomm, E., & Gerjets, P. (2020). How laypersons consider differences in sources' trustworthiness and expertise in their regulation and resolution of scientific conflicts. *International Journal of Science Education, Part B: Communication and Public Engagement, 10*(4), 335–354. https://doi.org/10.1080/21548455.2020.1849856

Hammer, D. (1995). Student inquiry in a physics class discussion. *Cognition and Instruction, 13*(3), 401–430. https://doi.org/10.1207/s1532690xci1303_3

Haskel-Ittah, M. (2023). Explanatory black boxes and mechanistic reasoning. *Journal of Research in Science Teaching, 60*(4), 915–933. https://doi.org/10.1002/tea.21817

Haskel-Ittah, M., & George-Levi, S. (2023). Characterisation of knowledge of cancer, illness perceptions and their interaction among high-school students. *International Journal of Science Education, 45*(9), 709–733. https://doi.org/10.1080/09500693.2023.2174819

Hendriks, F., & Kienhues, D. (2019). Science understanding between scientific literacy and trust: Contributions from psychological and educational research. In *Science communication* (pp. 29–50). De Gruyter. https://doi.org/10.1515/9783110255522-002

Joosten, E. A. G., DeFuentes-Merillas, L., De Weert, G. H., Sensky, T., Van Der Staak, C. P. F., & De Jong, C. A. J. (2008). Systematic review of the effects of shared decision-making on patient satisfaction, treatment adherence and health status. *Psychotherapy and Psychosomatics, 77*(4), 219–226. https://doi.org/10.1159/000126073

Keil, F. C., Stein, C., Webb, L., Billings, V. D., & Rozenblit, L. (2008). Discerning the division of cognitive labor: An emerging understanding of how knowledge is clustered in other minds. *Cognitive Science, 32*(2), 259–300. https://doi.org/10.1080/03640210701863339

Koslowski, B. (1996). *Theory and evidence: The development of scientific reasoning*. https://books.google.com/books?hl=iw&lr=&id=TbpYNhopCGoC&oi=fnd&pg=PR9&dq=Theory+and+evidence:+The+development+of+scientific+reasoning.&ots=UbG_dFQo3Q&sig=cuDmeZWkJ6gMk8iF2EHse69GbV4

Krist, C., Schwarz, C. V., & Reiser, B. J. (2019). Identifying essential epistemic heuristics for guiding mechanistic reasoning in science learning. *Journal of the Learning Sciences, 28*(2), 160–205. https://doi.org/10.1080/10508406.2018.1510404

Lee, V. R., & Dubovi, I. (2020). At home with data: Family engagements with data involved in type 1 diabetes management. *Journal of the Learning Sciences, 29*(1), 11–31. https://doi.org/10.1080/10508406.2019.1666011

Lin, Y., Docherty, S. L., Porter, L. S., & Bailey, D. E. (2020). Common and co-occurring symptoms experienced by patients with gastric cancer. *Oncology Nursing Forum, 47*(2), 187–202. https://doi.org/10.1188/20.ONF.187-202

Montori, V. M., Gafni, A., & Charles, C. (2006). A shared treatment decision-making approach between patients with chronic conditions and their clinicians: The case of diabetes. *Health Expectations, 9*(1), 25–36. https://doi.org/10.1111/j.1369-7625.2006.00359.x

National Academies of Sciences, Engineering, & and Medicine. (2016, November 14). *Science literacy: Concepts; contexts; and consequences.* Science Literacy: Concepts, Contexts, and Consequences; National Academies Press. https://doi.org/10.17226/23595

National Research Council. (2012). A framework for K-12 science education: Practices, crosscutting concepts, and core ideas. In *A framework for K-12 science education: Practices, crosscutting concepts, and core ideas*. https://doi.org/10.17226/13165

National Research Council. (2013). Next generation science standards: For states, by states. In *Next generation science standards: For states, by states* (Vol. 1–2). National Academies Press. https://doi.org/10.17226/18290

Nemirovsky, R. (2011). Episodic feelings and transfer of learning. *Journal of the Learning Sciences, 20*(2), 308–337. https://doi.org/10.1080/10508406.2011.528316

O'Connor, A. M. (1995). *User manual-decision self-efficacy scale*. Ottawa Hospital Research Institute. https://scholar.google.com/scholar?hl=iw&as_sdt=0%2C5&q=O'connor+A.+User+manual-decision+self-efficacy+scale.+Ottawa%3A+Ottawa+Hospital+Research+Institute+1995.&btnG=

Osborne, J., & Pimentel, D. (2023). Science education in an age of misinformation. *Science Education, 107*(3), 553–571. https://doi.org/10.1002/sce.21790

Punnett, G., Eastwood, C., Green, L., & Yorke, J. (2024). A systematic review of the effectiveness of decision making interventions on increasing perceptions of shared decision making occurring in advanced cancer consultations. *Patient Education and Counseling, 123*. https://doi.org/10.1016/j.pec.2024.108235

Rozenblum, Y., Dalyot, K., & Baram-Tsabari, A. (2024). People who have more science education rely less on misinformation—Even if they do not necessarily follow the health recommendations. *Journal of Research in Science Teaching*. https://doi.org/10.1002/TEA.21975

Russ, R. S., Scherr, R. E., Hammer, D., & Mikeska, J. (2008). Recognizing mechanistic reasoning in student scientific inquiry: A framework for discourse analysis developed from philosophy of science. *Science Education, 92*(3), 499–525. https://doi.org/10.1002/sce.20264

Ryder, J. (2001). Identifying science understanding for functional scientific literacy. *Studies in Science Education, 36*(1), 1–44. https://doi.org/10.1080/03057260108560166

Scalco, K. C., Talanquer, V., Kiill, K. B., & Cordeiro, M. R. (2018). Making sense of phenomena from sequential images versus illustrated text. *Journal of Chemical Education, 95*(3), 347–354. https://doi.org/10.1021/acs.jchemed.7b00716

Sevian, H., Hugi-Cleary, D., Ngai, C., Wanjiku, F., & Baldoria, J. M. (2018). Comparison of learning in two context-based university chemistry classes. *International Journal of Science Education, 40*(10), 1239–1262. https://doi.org/10.1080/09500693.2018.1470353

Sharon, A. J., & Baram-Tsabari, A. (2020). Can science literacy help individuals identify misinformation in everyday life? *Science Education, 104*(5), 873–894. https://doi.org/10.1002/sce.21581

Shauli, S., & Baram-Tsabari, A. (2019). The usefulness of science knowledge for parents of hearing-impaired children. *Public Understanding of Science, 28*(1), 19–37. https://doi.org/10.1177/0963662518772503

Southard, K., Espindola, M. R., Zaepfel, S. D., & Bolger, M. S. (2017). Generative mechanistic explanation building in undergraduate molecular and cellular biology*. *International Journal of Science Education, 39*(13), 1795–1829. https://doi.org/10.1080/09500693.2017.1353713

Stadtler, M., & Bromme, R. (2004). Laypersons searching for medical information on the web: The role of metacognition. In *Proceedings of the 26th Annual Conference of the Cognitive Science Society* (p. 1638). http://www.cogsci.northwestern.edu/cogsci2004/ma/ma293.pdf

Tabak, I. (2018). *Functional scientific literacy disciplinary literacy meets multiple source use*. pp. 221–237.

Tabak, I., & Dubovi, I. (2023). What drives the public's use of data? The mediating role of trust in science and data literacy in functional scientific reasoning concerning COVID-19. *Science Education, 107*(5), 1071–1100. https://doi.org/10.1002/sce.21789

Tamhane, S., Rodriguez-Gutierrez, R., Hargraves, I., & Montori, V. M. (2015). Shared decision-making in diabetes care. *Current Diabetes Reports, 15*(12) Current Medicine Group LLC 1. https://doi.org/10.1007/s11892-015-0688-0

Taube, O., Ranney, M. A., Henn, L., & Kaiser, F. G. (2021). Increasing people's acceptance of anthropogenic climate change with scientific facts: Is mechanistic information more effective for environmentalists? *Journal of Environmental Psychology, 73*, 101549. https://doi.org/10.1016/j.jenvp.2021.101549

Trouche, E., Chuey, A., Lockhart, K. L., & Keil, F. C. (2017). Why teach how things work? Tracking the evolution of children's intuitions about complexity. In *CogSci 2017 – proceedings of the 39th annual meeting of the cognitive science society: Computational foundations of cognition* (pp. 3368–3373).

Valladares, L. (2021). Post-truth and education: STS vaccines to re-establish science in the public sphere. *Science and Education, 31*(5), 1311–1337. https://doi.org/10.1007/s11191-021-00293-0

Vo, H., Valentine, K. D., Barry, M. J., & Sepucha, K. R. (2023). Evaluation of the shared decision-making process scale in cancer screening and medication decisions. *Patient Education and Counseling, 108*, 107617. https://doi.org/10.1016/j.pec.2022.107617

Wilensky, U., & Resnick, M. (1999). Thinking in levels: A dynamic systems approach to making sense of the world. *Journal of Science Education and Technology, 8*(1), 3–19. https://doi.org/10.1023/A:1009421303064

MIX
Papier aus verantwortungsvollen Quellen
Paper from responsible sources
FSC® C105338

If you have any concerns about our products,
you can contact us on
ProductSafety@springernature.com

In case Publisher is established outside the EU,
the EU authorized representative is:
Springer Nature Customer Service Center GmbH
Europaplatz 3, 69115 Heidelberg, Germany

Printed by Libri Plureos GmbH
in Hamburg, Germany